All About LIGHTNING

Martin A. Uman

Department of Electrical Engineering
University of Florida, Gainesville

Dover Publications, Inc., New York

Published in Canada by General Publishing Company, Ltd., 30 Lesmill Road, Don Mills, Toronto, Ontario.

Published in the United Kingdom by Constable and Company, Ltd.

This Dover edition, first published in 1986, is an unabridged and corrected republication of *Understanding Lightning,* originally published by Bek Technical Publications Inc., Carnegie, Pennsylvania, in 1971. The color illustrations of the original edition (frontispiece and pages 51, 52, 58, 97, 98, 99, 154 and 157) are reproduced in black and white in this Dover edition, but the frontispiece and the illustrations on pages 98 and 99 also appear in color on the covers.

Manufactured in the United States of America
Dover Publications, Inc., 31 East 2nd Street, Mineola, N.Y. 11501

Library of Congress Cataloging-in-Publication Data

Uman, Martin A.
    All about lightning.

    Unabridged and corrected republication of: Understanding lightning.
    Bibliography: p.
    Includes index.
    1. Lightning. I. Title.
QC966.U38   1986   551'.63   86-16766
ISBN 0-486-25237-X

All About
LIGHTNING

A lightning flash to a 23-foot-tall European ash tree near Lugano, Switzerland. The tree suffered no external damage. Photograph was taken from a distance of about 200 feet by Richard E. Orville, State University of New York at Albany, using Kodachrome II 35 mm daylight film, a lens setting of f/5.6, and the shutter on time exposure. (Reproduced in color on front cover.) Additional details are found in Orville, R. E., Photograph of a Close Lightning Flash, *Science, 162,* 666-667, 1968. Lightning photography is discussed in Chapter 10; lightning damage to trees in Chapter 5.

To my wife Dorit,

*who compiled the original list of questions
from which this book evolved,*

and to Mara, Jon, and Derek,

*who share with their father the fun of
watching lightning and listening to thunder.*

preface to the 1986 edition

All About Lightning (originally published under the title *Understanding Lightning*) has been revised for the Dover edition primarily in that some references have been updated and additional references have been added at the ends of some chapters. Although there have been significant advances in our understanding of the details of lightning since the first edition, the underlying fundamentals discussed in *All About Lightning* have not changed. Those who wish to continue with a technical book on lightning may be referred to my book *Lightning*. Originally published by McGraw-Hill in 1969, it was reprinted by Dover in 1984 with a new appendix summarizing the latest research up to that time.

Martin A. Uman
Gainesville, 1986

preface to the first edition

In lecturing about lightning to various groups of non-scientists ranging from high school students to Rotarians, I have found that the same questions are almost always asked. The identical questions are also posed by scientists who are not specialists in lightning. In *Understanding Lightning* these questions have been organized into groups which serve as chapter titles. Besides directly answering the questions in the chapters, I have incorporated in the answers — in non-technical terms — the important technical facts about lightning. Thus, by reading *Understanding Lightning* from beginning to end, the lay reader can acquire in a relatively painless fashion a good understanding of this fascinating natural phenomenon.

References to the source material discussed are included so that the reader may pursue his individual interests further. References are given for another reason. There is a great deal that scientists still do not understand about lightning, and hence there is room for opinion and speculation. Through the references the supporters of specific viewpoints are identified — information which should be of value to lightning researchers, particularly those just entering the field.

Most of my mail from high school and college science students includes a request for sources of information about ball lightning. Apparently quite a few science papers have been written on the subject. Because of this interest, the ball lightning chapter contains a particularly detailed bibliography.

In organizing the book in terms of questions and answers, it was sometimes necessary for one topic (the lightning current, for example) to be considered in more than one chapter. In order to find all references to a given topic, the reader must use the index. Every possible effort has been made to develop an index that is comprehensive and easy to use.

At the time this book was being written there were no non-technical books on lightning in print in English. In fact, only one moderately up-to-date layman's book on lightning is available in libraries: *The Lightning Book,* Peter E. Viemeister (Doubleday, 1961). Three technical books are available: *Lightning,* Martin A. Uman (McGraw-Hill, 1969), *The Flight of Thunderbolts*, Second Edition, B. F. J. Schonland (Clarendon Press, 1964), and *Physics of Lightning*, D. J. Malan (The English Universities Press Ltd., 1963). Schonland's book is the least technical of the three: much of it can be read and understood by the layman. Mine is the most technical and is the only existing book describing the important lightning research of the 1960s.

I wish to express my appreciation to Mr. Alan R. Taylor of the U.S. Forest Service for his assistance with Chapter 5 and for allowing me to use his unpublished statistics on lightning deaths and injuries in Chapter 3; and to Mr. Roderick N. Robinson of Bek Technical Publications for his valuable editing of the manuscript. I am indebted to many people at the Westinghouse Research Laboratories for their help in making this book possible. In particular, special thanks are due to the personnel of the Drafting and Graphic Arts departments, to Mrs. Martha Fischer who typed and retyped the manuscript, and to Dr. A. M. Sletten who read and criticized the manuscript. I am grateful to the various individuals and organizations who allowed me to use their excellent photographs. Credits are included in the captions.

My own research in lightning during the past ten years has been supported primarily by the Office of Naval Research. In particular, I owe a debt of gratitude to Mr. James Hughes of the Atmospheric Science Program, Office of Naval Research for his continuing interest in and support of my lightning studies. Without that interest and support this book would not have been written.

Martin A. Uman
Pittsburgh, 1971

contents

All About
LIGHTNING

chapter
one

Why Did Benjamin Franklin Fly the Kite?

Shuffle across a nylon rug on a dry winter day and your body acquires an excess electrical charge. That charge may be violently released by a spark jumping from your finger tip to a light switch or doorknob. Lightning is nothing more than a very long spark which discharges regions of excess electrical charge developed in thunderclouds. Frictional charging, that is charging by rubbing together certain dissimilar materials such as a shoe sole and the rug, has probably always been familiar to man. In 1746 Benjamin Franklin began his experiments in electricity. His experiments were made possible by (1) the frictional charging mechanism and (2) the fortuitous invention earlier that year of the Leyden jar — a primitive capacitor to store electrical charge.

Prior to Franklin's interest with electricity, a number of scientists had suggested that lightning might be an electrical phenomenon. Their concern, however, had gone no further than suggestion. In November 1749 Franklin wrote the following about the sparks (in his terminology, electrical fluid) he had studied.

Electrical fluid agrees with lightning in these particulars. 1. Giving light. 2. Colour of the light. 3. Crooked direction. 4. Swift motion. 5. Being conducted by metals. 6. Crack or noise in exploding. 7. Subsisting in water or ice. 8. Rending bodies as it passes through. 9. Destroying animals. 10. Melting metals. 11. Firing inflammable substances. 12. Sulphureous smell. The electrical fluid is attracted by points. We do not know whether this property is in lightning. But since they agree in all particulars wherein we can already compare them, is it not possible they agree likewise in this? Let the experiment be made. (Ref. 1.1)

Franklin was the first to design an experiment to prove that lightning was electrical. In July 1750 he wrote:

To determine the question whether the clouds that contain lightning are electrified or not, I would propose an experiment to be tried where it may be done conveniently. On the top of some high tower or steeple place a kind of sentry box . . . big enough to contain a man and an electrical stand [an insulator]. From the middle of the stand let an iron rod rise and pass bending out of the door, and then upright twenty or thirty feet, pointed very sharp at the end. If the electrical stand be kept clean and dry, a man standing on it when such clouds are passing low might be electrified and afford sparks, the rod drawing fire to him from the cloud. If any danger to the man should be apprehended (though I think there would be none), let him stand on the floor of his box and now and then bring near to the rod the loop of a wire that has one end fastened to the leads, he holding it by a wax handle; so the sparks, if the rod is electrified, will strike from the rod to the wire and not affect him. (Ref. 1.2)

His experiment and the results he expected to achieve are illustrated in Fig. 1.1. The aim was to show that the clouds were electrically charged, for if this were the case, it followed that lightning was also electrical. Franklin did not appreciate the danger involved in his experiment. If the iron rod were directly struck by lightning, the experimenter would almost certainly be killed. Such was eventually to be the case as we shall see.

In France in May 1752 Thomas-Francois D'Alibard successfully performed Franklin's suggested experiment. Sparks were observed to jump from the iron rod during a thunderstorm. It was proved that thunderclouds contain electrical charge. Soon after, the experiment was successfully repeated in France again, in England, and in Belgium. In July 1753 G. W. Richmann, a Swedish physicist working in

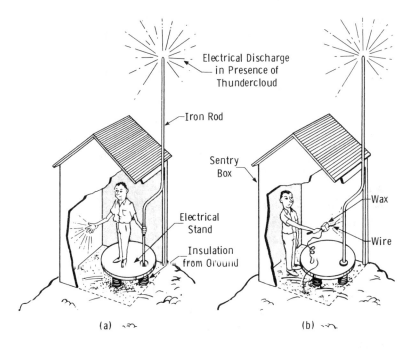

Fig. 1.1. Franklin's original experiment to show that thunderclouds are electrified. (a) Man on electrical stand holds iron rod with one hand and obtains an electrical discharge between the other hand and ground. (b) Man on ground draws sparks between iron rod and a grounded wire held by an insulating wax handle.

Russia, put up an experimental rod and was killed by a direct lightning strike.

Before Franklin himself got around to performing the experiment, he thought of a better way of proving his theory — an electrical kite. It was to take the place of the iron rod, since it could reach a greater elevation than the rod and could be flown anywhere. During a thunderstorm in 1752 Franklin flew the most famous kite in history. Sparks jumped from a key tied to the bottom of the kite string to the knuckles of his hand (Fig. 1.2). He had verified his theory, and had probably done so before he knew that D'Alibard had already obtained the same proof.

There is some controversy as to whether Franklin flew his kite in June 1752 or later that summer and whether at the time of his experiment he knew of the earlier French results (Refs. 1.3, 1.4). Curiously, it wasn't until 1788 that Franklin himself first wrote that he had performed the kite experiment; and then only a brief sentence was devoted to the subject (Ref. 1.5). Nevertheless, in October 1753 Franklin described the kite experiment in detail and stated that it had succeeded in Philadelphia — but not that he himself had performed it (Ref. 1.6). The classic account of Franklin's kite flight was written by Joseph Priestly in his *History and Present State of Electricity* published 15 years after the flight (Ref. 1.7). Evidence is available to show that Franklin had read Priestly's manuscript before publication and had approved of it (Ref. 1.3).

Kite flying in thunderstorm weather can be dangerous. A number of people have been killed imitating Benjamin Franklin. In the nineteenth and early twentieth century meteorological observations were made by sending instruments aloft on large box kites. In 1898, 17 U.S. Weather Bureau stations were equipped for daily kite ascensions. The kites used weighed 8 lb, carried 2 lb of instruments, and dragged as much as 20 or 30 lb of piano wire beneath them.

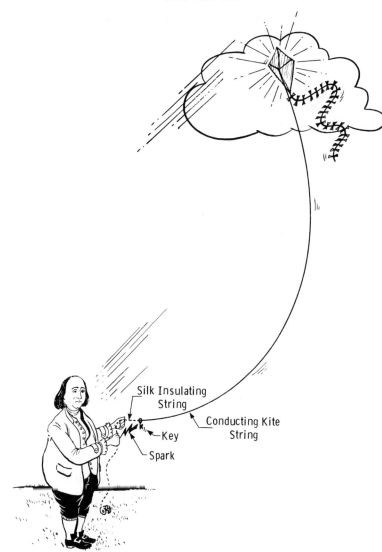

Fig. 1.2. Franklin's electrical kite experiment: Sparks jump from the electrified key at the end of the electrified kite string to Franklin's hand.

When a kite was struck by lightning the piano wire vaporized in a bright flash. After the flash, the remains of the piano wire were briefly evident as rusty smoke.

Details of two cases of lightning strikes to meteorological kites are described in Ref. 1.8. In one case a man assisting in the flight was killed. Both times, roughly a mile of wire had run out when the strike occurred.

REFERENCES

1.1. Smyth, Albert H., *The Writings of Benjamin Franklin*, Vol. 3 of 10 Vols., MacMillan, New York, 1905-1907, p. 255.
1.2. Smyth, Albert H., ibid, Vol. 2, pp. 434-438.
1.3. Jernegan, Marcus W., Benjamin Franklin's "Electrical Kite" and Lightning Rod, *New England Quarterly, 1*, 180-196 (1928).
1.4. Van Doren, Carl, *Benjamin Franklin*, The Viking Press, New York, 1938, pp. 156-173.
1.5. *Benjamin Franklin: The Autobiography and Other Writings*, Signet Paperback, The New American Library, New York, 1961, p. 167.
1.6. Smyth, Albert H., ibid, Vol. 3, pp. 99-100.
1.7. *Benjamin Franklin: The Autobiography and Other Writings*, Signet Paperback, The New American Library, New York, 1961, pp. 234-235 gives Priestly's account of Franklin's kite flight.
1.8. Brunk, I. W., Ben Franklin Was Lucky, *Weatherwise, 11*, 92-93, June 1958; Lightning Death During Kite Flight (or Radiosondes are Safer), *Weatherwise, 11*, 204-205, December 1958.

chapter two

How Does a Lightning Rod Work?

It is a common misconception that lightning rods discharge clouds and thus prevent lightning. Actually lightning rods only serve to route the lightning harmlessly to ground. In doing so they divert the lightning when it is 10 to 100 yards away.

In 1749 Benjamin Franklin wrote a letter which was published in *Gentlemen's Magazine,* May 1750. It read, in part,

There is something however in the experiments of points, sending off or drawing on the electrical fire, which has not been fully explained, and which I intend to supply in my next . . . from what I have observed on experiments, I am of opinion that houses, ships, and even towers and churches may be eventually secured from the strokes of lightning by their means; for if instead of the round balls of wood or metal which are commonly placed on the tops of weathercocks, vanes, or spindles of churches, spires, or masts, there should be a rod of iron· eight or ten feet in length, sharpened gradually to a point like a needle, and gilt to prevent rusting, or divided into a number of points, which would be better, the electrical fire would, I think, be drawn out of a cloud silently, before it could come near enough to strike.

This is Franklin's earliest recorded suggestion of the lightning rod. In the "experiments of points" he placed electrical charge on isolated conductors and then showed that the charge could be drained away (discharged) slowly and silently if a pointed and grounded (attached to ground) conductor were introduced into the vicinity. When the pointed conductor was brought too close to the charged conductor, the discharge occurred violently via an electric spark.

In the discussion in which he proposed the original experiment to determine if lightning were electrical (July 1750 — see Chapter 1), Franklin repeated his suggestion for protective lightning rods, adding that they should be grounded (i.e., that a wire should connect the lightning rod to the ground or, in the case of a ship, to the water).

Lightning rods were apparently first used for protective purposes in 1752 in France and later the same year in the United States (Refs. 2.1, 2.2).

Franklin originally thought — erroneously — that the lightning rod silently discharged the electric charge in a thundercloud and thereby prevented lightning. However, in 1755 he stated:

I have mentioned in several of my letters, and except once, always in the *alternative, viz.,* that pointed rods erected on buildings, and communicating with the moist earth, would *either* prevent a stroke, or, if not prevented, would *conduct* it, so that the building should suffer no damage. (Ref. 2.3)

It is in the latter manner that lightning rods actually work. The charge flowing between a lightning rod and a thundercloud is much too small to discharge the thundercloud (Ref. 2.4). The rod diverts to itself a stroke on its way to earth but can do so only in the final part of the stroke's earthward trajectory. Diversion is achieved by the initiation of an electrical discharge (Fig. 2.1), a sort of traveling spark, which propagates from the rod, intercepts the downward-moving lightning, and provides a conducting path to the rod. Before the traveling spark is initiated, the downward-moving lightning is essentially uninfluenced by objects on the ground beneath it. The traveling spark is generally 10 to 100 yards long when it meets the lightning.

Any high object may initiate an upward-moving spark which attempts to reach the downward-moving lightning. It is therefore important that the lightning rod be the tallest object near the structure it protects, so that its traveling spark catches the lightning rather than a spark initiated by the chimney or a nearby tree.

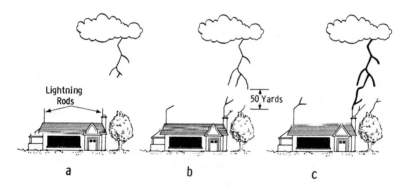

Fig. 2.1. Final stages of a strike to a lightning rod. The time interval between B and A and between C and B is about 1/5000 second. The lightning channel becomes very bright (view C) immediately after the upward-traveling spark connects with the downward-moving lightning. Drawings are not to scale.

Franklin refused to patent the lightning rod or otherwise to profit by its invention.

No lightning rod, however tall, can offer absolute protection; lightning has struck the Empire State Building 50 ft below the top (see Chapter 6). Nevertheless, lightning rod systems are effective if used properly, and many codes have been written to describe their use (Ref. 2.5). A single vertical rod will almost always protect objects within an imaginary cone (the "cone of protection") formed by all lines connecting the top of the rod at height H with a circle on the ground, beneath the rod, of radius between H and 2H. In Fig. 2.2a, the house has a single lightning rod with a cone of protection assumed to have a base radius H in accord with the British lightning code. To protect a large house it is more practical to use multiple lightning rods: in Fig. 2.2b, three cones of protection overlap to provide a large volume of protection without excessive height. The house in Fig. 2.1 is protected according to the U.S. lightning code which specifies a base radius of 2H for the cone of protection. The smaller the base radius, the greater the probability that no lightning strokes will violate the cone. Thus, a structure containing explosives or highly flammable materials is often protected by a cone with a base radius as small as H/2.

A lightning rod system has three main parts: the rods on the roof, the wires which connect the rods together and those which run down the sides of the house or building to the grounding arrangement, and the grounding arrangement. Although the rods are the most visible, each of the three parts is equally important since the system may fail if any part is inadequate. Any metal rod or pipe may be an effective lightning rod, but to ensure a long lifetime for the rod, corrosion-resistant metal such as copper, aluminum, or galvanized iron should be used. There is no evidence that a pointed rod is better than one with a ball on the top. A wide variety of lightning rod shapes can be seen on urban and rural structures.

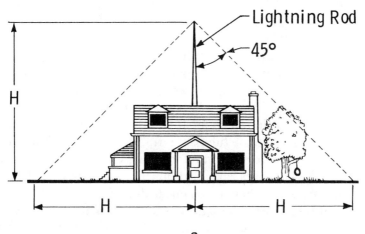

a

b

Fig. 2.2. (a) A house protected by a single lightning rod having an assumed 45°-angle cone of protection — that is, the height of the rod is H and the base area assumed to be safe from a lightning strike has a radius H. (b) The same house protected by a lightning rod system consisting of three smaller rods, each assumed to provide a 45°-angle cone of protection.

The primary function of the wires which link the lightning rods and those connecting the rods to the grounding arrangement is to carry the lightning current from the rods to the ground. The wires on the roof have the secondary function of intercepting lightning discharges which may have missed the rods. In fact, grids of wires alone have been used on roofs in place of lightning rods. The lightning codes recommend aluminum or copper wires. The U.S. lightning code suggests approximately 1/4 in. diameter copper wire or 3/8 in. diameter aluminum wire, while the British code recommends 3/8 in. diameter for both. In addition to round solid wire conductors, tubular, strip, or stranded aluminum or copper conductors of an appropriate cross-section may be used. The wire sizes specified in the codes appear to be chosen partly for their mechanical durability as well as for their ability to carry the lightning current. Wire several times thinner than that recommended will carry all but the most extreme lightning currents without damage (Refs. 2.7, 2.8). An example of a lightning protection system using small diameter wire is given later in this chapter.

Wires carrying the lightning current must be well grounded, otherwise the lightning may jump from the wires into the protected structure in search of a better ground. Grounding is best accomplished by connecting the wires to long rods which are driven into the ground or by connecting the wires to large buried metallic conductors. The rods or buried conductors should in turn be connected to all nearby gas pipes, water pipes, or other buried metallic pipes or cables.

It is sometimes imperative to keep the lightning current and possible attendant sparks from contacting any part of a protected structure — a typical case being a liquid-fuel storage vat in which flammable vapors are present. Here, the roof rods and wire conductors are often replaced by a system of wires suspended between tall towers arranged around the structure. A similar scheme is used to protect high voltage

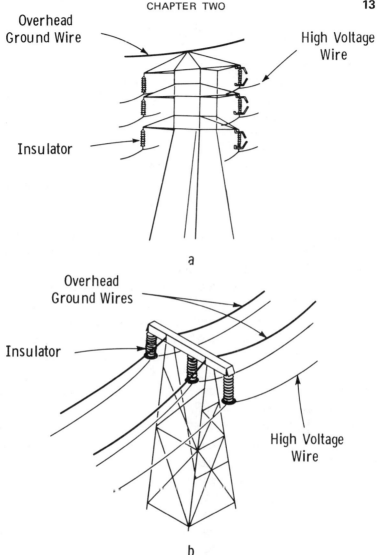

Overhead
Ground Wire

High Voltage
Wire

Insulator

a

Overhead
Ground Wires

Insulator

High Voltage
Wire

b

Fig. 2.3. Use of overhead ground wires to protect high voltage transmission lines from lightning. Each of the high voltage wires is supported by the bottom of an insulator. Ground wires are attached to the metal transmission line towers which are well grounded. Drawings are not to scale. (a) A typical 132,000 volt line. (b) A typical 220,000 volt line.

transmission lines from lightning strikes. A grounded wire (or wires) is strung above the high voltage lines to intercept strokes that would otherwise hit the power lines (Fig. 2.3).

The same principle has been adopted successfully in Poland for the protection of small farmhouses (Ref. 2.8). These buildings are usually made of wood and frequently have thatched roofs, thus making them very susceptible to lightning-caused fires. Since their inhabitants can rarely afford the expense (several hundred dollars) of a protective system satisfying the formal codes, Stanislas Szpor suggested a simpler, inexpensive system which the residents can install themselves. Basically, it consists of a 1/8 in. diameter galvanized iron wire suspended above the roof ridge from two small wooden towers installed at the ends of the roof ridge. From each tower the wire slopes downward and is buried in the ground. Over a five-year period, structures so protected had only about 10% of lightning-caused fires that unprotected structures suffered. Since about 2000 rural structures are ignited each year by lightning in the U.S. and since probably most are not protected because of the expense of the system specified by the U.S. code, it appears that Szpor's lightning protection system or some system similar to it could be advantageous to U.S. farmers.

Information about the number of lightning strikes per year to structures of various heights as well as to flat ground is given in Chapters 3, 6, and 7.

REFERENCES

2.1. Jernegan, Marcus W., Benjamin Franklin's "Electrical Kite" and Lightning Rod, *New England Quarterly, 1,* 180-196 (1928).

2.2. Van Doren, Carl, *Benjamin Franklin,* The Viking Press, New York, 1938, pp. 156-173.

2.3. Cohen, I. B., *Benjamin Franklin's Experiments,* Harvard University Press, Cambridge, Mass., 1941, letter to D'Alibard of June 29, 1755.

2.4. Golde, R. H., The Lightning Conductor, *J. Franklin Inst., 283,* 451-477 (1967).

2.5. The Protection of Structures against Lightning, British Standards Institution, 2 Park Street, London, 1965.
Lightning Protection Code 1980, National Fire Protection Association Publication 78, National Fire Protection Association, Inc., 470 Atlantic Avenue, Boston, Mass. 02210.
Recommendations pour les installations de protection contre la foudre, Association Suisse de Electriciens, Zürich, 1959.
Allgemeine Blitzschutz-Bestimmungen, Ausschuss für Blitzableiterbau e.V. (ABB), Berlin, 1963.

2.6. Golde, R. H., Protection of Structures against Lightning, *Proc. IEE, Control and Science, 115,* 1523-1529 (1968).

2.7. Müller-Hillebrand, D., The Protection of Houses by Lightning Conductors, *J. Franklin Inst., 274,* 34-54 (1962).

2.8 Szpor, S., Paratonnerres ruraux de type léger, *Revue Générale de l'Électricite, 68,* 263-270 (1959).

ADDITIONAL READING

Golde, R. H., *Lightning Protection,* Edward Arnold, London, 1973.

chapter three

How Many People Are Killed By Lightning Each Year?

What Should I Do If Caught Outdoors In A Thunderstorm?

Is It Safe To Talk On The Telephone During A Thunderstorm?

Should I Unplug Radios And TVs?

How Can I Help Someone Struck By Lightning?

Although no exact figures on lightning fatalities are available — there is no central agency to which lightning accidents must be reported — various studies (Refs. 3.1, 3.2, 3.3, 3.4) have placed the average annual number of deaths in the U.S. at 100, 137, 151, 300, and even 600. Whatever the exact number, lightning causes more direct deaths than any other weather phenomena. Snowstorms cause more indirect deaths (e.g., heart attacks from shoveling the driveway). Tornadoes cause about half as many deaths as lightning but inflict much more property damage. It is perhaps surprising that about two-thirds of the people involved in lightning accidents subsequently make a complete recovery. Most, if not all, of these survivors are probably not directly struck by lightning, but rather receive electrical shock from being in the close vicinity of a strike.

17

About 70% of all fatal lightning accidents involve one person, about 15% occur in groups of two, the remainder involves three or more deaths per lightning event. Between 75 and 85% of all lightning deaths and injuries are to men, probably because they are more likely to be out-of-doors. About 70% of all injuries and fatalities occur in the afternoon, about 20% between 6 p.m. and midnight (Ref. 3.1).

The largest single category of lightning deaths (12 to 15% of all fatalities) is composed of those unfortunate individuals who seek refuge under trees during thunderstorms and have their sheltering trees struck by lightning. Perhaps one-third of these are golfers. Two additional categories in which large numbers of lightning fatalities occur are designated "Open Water" and "Tractors", each claiming about 10% of the total fatalities. "Open Water" includes people fishing, swimming, on beaches, piers, and levees, in boats, and on water skis; "Tractors" includes people on, in, or near farm tractors or other implements, construction equipment, cars, and trucks (Refs. 3.1, 3.3).

Most lightning deaths occur outdoors. Over 30% of all lightning deaths involve people who work outdoors; over 25% of all deaths involve outdoor recreationists. A breakdown of deaths and injuries to farmers and ranchers and to outdoor recreationists for the 20-year period 1950 through 1969 is given in Table 3.1. The apparent increase from the 1950s to the 1960s in the number and percentage of deaths to outdoor recreationists is perhaps due to the increasing number of outdoor recreationists. Table 3.2 gives more detailed data on death and injury to outdoor recreationists, while Table 3.3 breaks down the "Open Water" category of Table 3.2.

In addition to humans, animals are killed by lightning. In 1968, lightning caused the death of 464 cattle (362 killed directly, the remainder by lightning-caused fires), 13 horses (11 killed directly), 42 hogs, 2 dogs, and 1 white ox resident in a museum pasture (Ref. 3.4).

TABLE 3.1. Deaths and injuries from lightning
in the 48 contiguous United States during 1950-1969 (Ref. 3.4)

(1)	All People		Outdoor Recreationists				Farmers and Ranchers				Unknown				Other Known			
	(2)†	(3)‡	Number		Percent		Number		Percent		Number		Percent		Number		Percent	
			(4)	(5) In-	(6)	(7) In-	(8)	(9) In-	(10)	(11) In-	(12)	(13) In-	(14)	(15) In-	(16)	(17) In-	(18)	(19) In-
Year	Killed	jured	Killed	jured	Killed	jured	Killed	jured	Killed	jured	Killed	jured	Killed	jured	Killed	jured	Killed	jured
1950	59	78	10	15	16.9	19.2	12	11	20.3	14.1	29	17	49.2	21.8	8	34	13.6	44.8
1951	95	109	9	17	9.5	15.6	28	12	29.5	11.0	41	33	43.2	30.3	17	47	17.9	43.1
1952	88	94	7	13	8.0	13.8	27	16	30.7	17.0	30	42	34.1	44.7	24	23	27.3	24.5
1953	71	98	8	13	11.3	13.3	9	4	12.7	4.8	44	56	62.0	57.2	10	25	14.1	25.5
1954	113	203	28	39	24.8	19.2	26	32	23.0	15.8	50	82	44.3	40.4	9	50	8.0	24.6
1955	113	206	11	15	9.7	7.3	38	56	33.6	27.3	51	93	45.2	45.2	13	42	11.5	20.4
1956	104	237	19	58	18.3	24.4	40	50	38.5	21.1	33	65	31.7	27.4	12	64	11.5	27.0
1957	112	277	23	38	20.5	13.7	29	40	25.9	14.4	43	67	38.4	24.2	17	132	15.2	47.7
1958	73	243	23	53	31.5	21.8	15	48	20.6	19.8	26	77	35.6	31.7	9	65	12.3	26.8
1959	153	326	39	56	25.4	17.2	38	51	24.8	15.6	47	125	30.7	38.4	29	94	18.9	28.8
1960	100	199	21	44	21.0	22.1	28	27	28.0	13.6	32	70	32.0	35.2	19	58	19.0	29.1
1961	115	254	24	33	20.9	13.0	34	38	29.6	15.0	40	92	34.8	36.2	17	91	14.8	35.8
1962	122	243	36	65	29.5	26.7	37	38	30.3	15.6	29	80	23.8	32.9	20	60	16.4	24.7
*1963	136	209	35	44	25.8	21.0	42	68	30.9	32.6	40	59	29.4	28.2	19	38	14.0	18.2
1964	108	241	22	42	20.4	17.4	35	47	32.4	19.5	20	79	18.5	32.7	31	73	28.7	30.3
1965	126	231	39	60	31.0	26.0	38	42	30.2	18.2	33	54	26.2	23.4	16	75	12.7	32.5
1966	88	188	32	76	36.4	40.4	21	27	23.8	14.4	13	64	14.8	34.0	22	21	25.0	11.2
1967	73	157	31	53	42.4	33.7	17	24	23.3	15.3	7	34	9.6	21.6	18	46	24.7	29.3
1968	108	363	42	149	38.9	41.1	38	36	35.2	9.9	11	48	10.2	13.2	17	130	15.7	35.8
1969	97	200	35	58	36.1	29.0	35	47	36.1	23.5	9	31	9.3	15.5	18	64	18.6	32.0
	2054	4156	494	941			587	714			628	1268			345	1233		

* Totals in 1963 do *not* include 81 fatalities in airliner crash apparently caused by lightning (Chapter 4).
† Column 2 equals the sum of Columns 4, 8, 12 and 16. Column 3 equals the sum of Columns 5, 9, 13 and 17.

TABLE 3.2. Known lightning deaths and injuries among outdoor recreationists in the 48 contiguous United States during 1950-1969 (Ref. 3.4)

Activity at time of strike	Killed	Injured
In, on, or near open water	200	177
Golfing	95	164
Camping or picnicking	36	188
Viewing or participating in athletics	29	65
Hiking or climbing	19	45
In, on, or near land vehicle	7	8
Horseback riding (pleasure)	9	8
Other recreation*	65	190
Unknown recreation**	34	96
Total	494	941

*Includes sightseers, recreation strollers, youths at play (except athletics) anywhere outdoors, and others.

**Persons identified by data source as having been at recreation areas at time they were struck, with no specific activity indicated.

TABLE 3.3. Analysis of "In, on, or near open water" in Table 3.2 (Ref. 3.4)

Sub-Category	Killed	Injured
Fishing from shore, bank	39	30
Fishing from boat	33	21
Other boat	30	38
Swimming	9	9
Other and unknown water*	89	79
Total	200	177

*Mostly persons on lake or ocean shores or stream banks and persons whose activities were listed in the data sources only as "fishing" with no indication as to whether they were on shore or in a boat.

From the foregoing it is obvious what you *should not do* when out-of-doors during a thunderstorm: do not make a lightning rod of yourself and do not stand beneath a lightning rod. Avoid projecting above the surrounding landscape, as you would do if you were standing in an open field, on the beach, or fishing from a small boat. Avoid standing under isolated trees or under any other isolated shelters. (Potentially dangerous sun and rain shelters are often provided on golf courses.) Stay away from wire fences, overground pipes, rails, and other metallic paths along the ground which could carry lightning currents to you from a stroke which has hit some distance away. Avoid swimming, since the current from a nearby stroke to the water can flow through the water to you.

Here's what you *should do* if caught out-of-doors during a thunderstorm. In an urban area, seek shelter in a building (preferably with metal frame and/or with a lightning rod system) or in a car (not a convertible) with the windows rolled up (see Chapter 4). In the wide-open spaces, find a ravine, valley, or, as a last resort, a depression in the ground. Crouch or lie down. In a wooded area, seek shelter in dense woods or among a thick growth of small trees.

Relatively few people are killed indoors by lightning. The greatest number of indoor deaths is probably due to lightning-caused fires. Nevertheless, between 1959 and 1965 four people were reported killed by lightning while using their telephones (Ref 3 1); many others were killed because they were near, or in contact with, fixtures connected to the house plumbing or to appliances that were plugged into the house wiring system.

When lightning strikes a house without a lightning rod system, the lightning current generally follows the water pipes and electrical wiring in an attempt to get to ground. It may also enter a house on the telephone or electric wires or on the lead-in wires from an exterior TV antenna. (The mast to which an external TV antenna is attached, and from which it is electrically isolated, should be well connected to ground in accord with the codes for lightning rod systems.) If the particular conductor in which the current is traveling is not well-connected to ground, the lightning may choose to leave that conductor and jump through the air to what it regards as a better path to ground. The trip through the air, or *side flash,* accounts for a significant fraction of indoor lightning fatalities (as well as being one of the ways in which people standing beneath lightning struck trees can be killed). Thus during a thunderstorm it makes good sense to stay far away from bath tubs, sinks, plugged-in appliances, telephones, or any objects that have a metallic connection with the outside.

Telephone and electric companies attempt to design systems which discourage lightning from entering houses via their wires. A "lightning protector" (costing about $50, excluding installation) on the main circuit breaker or fuse box will short-circuit to ground lightning current attempting to enter a house on the electric wires. A similar "lightning protector" is sold for TV lead-in wires. Protection of your house against direct lightning strikes can be accomplished with a lightning rod system (Chapter 2) at a cost of several hundred dollars (including installation). If lightning does enter electrical wiring, it can cause considerable damage to plugged-in appliances. Although it is safest to unplug all appliances before a thunderstorm, this is somewhat impractical. For a typical urban house in a region of moderate thunderstorm activity, say Pennsylvania or New York (see Chapter 7), the frequency of strikes to the house is statistically about once every hundred years (see Chapters 6 and 7). Saying the same thing another way, one in every hundred houses is hit each year.

Lightning "deaths" are often reversible. Many victims who appear dead, in that they are not breathing and have no heart beat, can be revived with proper first aid (Ref. 3.5). It is tragic that this fact is not more widely known. If a victim is still breathing, he will, in all probability, recover.

Often, when a person is involved in a lightning accident, heart action and breathing stop instantly. There is medical evidence that all metabolism seems to stop (Ref. 3.5). Typically the heart starts again, but respiration does not. If the victim does not breathe for a prolonged period, he suffers brain damage from lack of oxygen. The apparent stopping or slowing of metabolism can seemingly lengthen the usual five- or six-minute period before appreciable brain damage occurs. A case was reported in which the victim apparently did not breathe for between 13 and 22 minutes, and still made an essentially complete recovery (Ref. 3.6).

If the victim's heart has started spontaneously but his breathing has not, his airway should first be cleared and then mouth-to-mouth artificial respiration given at about one breath every 5 seconds. If his heart is not beating (if the victim has no pulse), both heart action and breathing must be restarted. Heart action can be stimulated by placing the victim on his back and pressing firmly on his chest (on the lower half of the sternum above the point where the ribs come together) with the heel of the hand once every second or slightly faster. A person alone with the victim should alternate between 15 chest compressions and 2 mouth-to-mouth respiration cycles. If two rescuers are present, one should compress the chest at a steady rate of once per second while the other gives one breath every five compressions timed so that the breath is going in as the hands are coming up. This first aid should be continued until the victim's heart action and breathing begin or until professional medical help is secured. The above is only a brief explanation of the rudiments of cardio-pulmonary resuscitation (CPR) and is no substitute for a proper education in these life saving procedures (Ref. 3.7).

REFERENCES

3.1. Zegel, F. H., Lightning Deaths in the United States: A Seven Year Survey from 1959 to 1965, *Weatherwise, 20*, 168-173, 179, August 1967.
3.2. U.S. Dept. of Commerce, ESSA, *Lightning,* 1966, U.S. Government Printing Office, Washington, D. C.
3.3. Metropolitan Life Insurance Company, *Statistical Bulletin, 33,* 8-10, June 1952.
3.4. Unpublished analysis by Alan R. Taylor, Northern Forest Fire Laboratory, USDA Forest Service, Missoula, Montana. Analysis based on data published in *Climatological Data, National Summary,* U.S. Department of Commerce, Weather Bureau, Vols. 1-9, 1950-58, and *Storm Data,* U.S. Department of Commerce, ESSA, Vols. 1-11, 1959-69.
3.5. Taussig, H. B., "Death" from Lightning and the Possibility of Living Again, *American Scientist, 57* (No. 3), 306-316 (1969).
3.6. Ravitch, M. M., R. Lane, P. Safar, F. M. Steichen, and P. Knowles, Lightning Stroke, *New England Journal of Medicine, 264,* 36-38 (1961).
3.7   Red Cross, *Red Cross CPR Module,* The American Red Cross, 1980. '

ADDITIONAL READING

Golde, R. H., and Lee, W. R., Death by Lightning, *Proc. IEE, 123,* 1163-1179, October 1976.
Cooper, M. A., Lightning Injuries: Prognostic Signs for Death, *Ann. Emerg. Med., 9,* 134-138, March 1980.
Storm Data, Volume 25, No. 12, December 1983, available from Publications Section (E/CC413), the National Climatic Data Center, NOAA, Federal Building, Asheville, N.C. 28801-2696. Contains lightning death and injury statistics by month and by state for the United States from 1959 to 1983.

chapter four

Am I Safe From Lightning In An Airplane? In A Car?

To be safe from lightning really means being certain that none of the lightning current can flow through you. If lightning is occurring nearby, this can only be ensured by providing the lightning current with paths to ground more preferred (paths which provide much less resistance) than your body. Lightning does not respect token levels of insulation. It is therefore suicidal to suppose that, for instance, the wearing of rubber shoes will protect you from a direct lightning strike. Lightning which travels many miles through insulating air is not about to be halted by half an inch or even a yard of insulating rubber. The rubber tires on a car *do not* serve to insulate the car from ground and thereby prevent it from being struck by lightning, as is commonly believed. A car is a relatively safe shelter in a thunderstorm because, if it is struck by lightning, the current will tend to flow in the metal skin of the car and not in the occupant. The lightning will ground itself by jumping from the car to the earth either through the air, along the surface of a tire, or through a tire. In the latter case, the tire will be destroyed. For maximum security, the car windows should be rolled up and wet (so that they conduct current along their surfaces), and the occupant should not touch any metal part of the car or the car radio.

Because modern commercial aircraft are essentially all metal, lightning currents seldom penetrate them. An airplane *on the ground* is a relatively safe refuge during a thunderstorm. What about planes in the air? Are aircraft in flight hit by lightning? If hit, do they continue to fly?

Airplanes in flight *are* struck by lightning. Between January 1965 and December 1966, the Federal Aviation Administration (FAA) collected reports of about 1000 lightning strikes to commercial aircraft. On the average, a given commercial airplane is struck by lightning once in every 5000 to 10,000 hours of flying time (Refs. 4.1, 4.2). As extreme examples of lightning activity, on one day in January 1969 four airliners were hit by lightning in separate incidents near Los Angeles (Ref. 4.3), and on September 26, 1964 a Boeing 727 in a holding pattern at Chicago was hit by lightning five separate times within a 20-minute period (Ref. 4.4).

Aircraft struck by lightning almost always continue to fly. Generally, lightning leaves pit marks or burn marks on the aircraft's metallic skin or burn or puncture holes through it. The FAA reports hole diameters up to 4 in., a common size being 1/2 in. (Ref. 4.1). Photographs of a lightning-damaged aircraft wing are shown in Figs. 4.1, 4.2, and 4.3. In addition to the generally minor damage caused to metallic aircraft parts, unprotected non-metallic parts may be more severely damaged. (Differences between lightning damage to metallic and non-metallic materials are discussed in the next chapter.)

Fig. 4.1. Left wing tip of a Boeing 707 which exploded after being hit by lightning near Elkton, Maryland on December 8, 1963. On the model insert the white portion of the wing gives the size and orientation of the portion of the actual wing shown. A number of lightning-caused holes and considerable pitting are evident. (Courtesy, Bernard Vonnegut and Roger Cheng, State University of New York at Albany)

Figure 4.4 shows lightning damage to the plastic nose section or radome of a commercial jetliner. (The radome is necessarily non-metallic so that the plane's radar can operate through it — radar signals are reflected by metal.) The aircraft in Fig. 4.4 had its radar unit put out of commission. Lightning is clearly a potential hazard. On the other hand, thunderstorm turbulence is a greater hazard. Severe turbulence can cause structural failure of the aircraft or loss of control by the pilot. The problem of loss of control is particularly critical during take-off and landing.

Lightning is a danger to aircraft for several reasons. It can fuse, burn away, or otherwise destroy mechanical or electrical parts; it can temporarily blind the pilot causing him to lose control of the aircraft; and it can ignite the aircraft's fuel. Fuel ignition was the probable cause of the only two

Fig. 4.2. Closer view of the Boeing 707 wing tip shown in Figure 4.1. Five lightning-caused holes are visible. (Courtesy, Bernard Vonnegut and Roger Cheng)

Fig. 4.3. Close-up of the major wing-tip hole and surrounding pitting shown at the left of Figures 4.1' and 4.2. (Courtesy, Bernard Vonnegut and Roger Cheng)

acknowledged lightning-related disasters involving modern commercial aircraft.

On June 26, 1959 near Milan, Italy, a Lockheed Constellation, a propeller-driven U.S. air carrier, was climbing in thunderstorm activity through about 11,000 ft when structural disintegration occurred. All persons aboard were killed. Investigation revealed that two of the fuel tanks had exploded. Fuel vent pipes extending off the trailing edges of the wings were exhausting air and fuel vapors (to maintain a pressure inside the fuel tank equal to the decreasing pressure outside, as the plane climbed) and were a likely point for fuel ignition. It is not clear whether lightning struck a fuel vent pipe or whether a traveling spark (similar to those which leave lightning rods; Chapter 2) was initiated at the pipe causing ignition. In either event, it is probable that some sort of electric spark ignited the fuel at the vent pipe, the burning fuel propagated into the number 6 and 7 fuel tanks, and those fuel tanks exploded (Ref. 4.1).

On December 8, 1963, a Boeing 707 was in a holding pattern at 5000 ft near Elkton, Maryland. There was thunderstorm activity in the area. Witnesses reported a lightning stroke near the aircraft concurrent with or immediately followed by explosion and burning of the 707. All aboard were killed. It was found that three fuel tanks had exploded and that there were lightning strike marks and holes on the left wing tip (Figs. 4.1, 4.2, 4.3). Exactly how the fuel tanks were ignited remains unknown. The fuel vent outlets on the 707 are underneath the wings and are recessed into them, so it is unlikely that lightning could strike the outlets directly or that they could be the source of traveling sparks. Further, when recovered, the vent outlets showed no signs of damage. Possibly a lightning stroke burned through the wing surface into a fuel tank (the fuel tank container is the wing skin). No evidence for this effect was found, although the explosion might have destroyed any pertinent evidence. The

Fig. 4.4. Lightning damage to the radome of a commercial jetliner —
one of four planes hit by lightning on one day in January 1969 in
separate incidents near Los Angeles (Ref. 4.3). ("Los Angeles Times"
photograph)

favored explanation of how the explosion occurred is that lightning caused sparks at some point where an explosive air-fuel mixture was present, perhaps in a fuel tank or near a fuel leak (Refs. 4.1, 4.2, 4.5).

The explosions of both the Lockheed Constellation and the Boeing 707 would not have occurred had the fuel tanks not contained an explosive mixture of oxygen and fuel vapor above the liquid fuel. In the absence of oxygen, the fuel can neither burn nor explode. At the time this book is being written, systems for replacing the oxygen in an aircraft's fuel tanks with nitrogen are being tested (Ref. 4.2). Such a system would make aircraft immune to explosions from lightning as well as from fires due to accidents in which the fuel is not exposed to the oxygen of the air (accidents in which the fuel tanks are not ruptured).

In addition to these two commercial aircraft, a number of military aircraft have also been destroyed by lightning. No official count is available, but it is known that, for example, in April 1970 an Air Force jet exploded after being hit by lightning, that there have been several recent cases of lightning-induced electrical failures on jet aircraft in flight necessitating crew ejection, and that eight servicemen were killed in a recent helicopter crash when lightning hit and damaged the rotor blades. Engines on many jets have been temporarily extinguished by lightning strikes, and in July 1969 a lightning strike to a jet resulted in two bombs being released.

When lightning strikes a plane, the plane substitutes for part of the lightning's air path. Typically lightning attaches to the plane's extremities, often entering at one wing tip and leaving by the top of the tail. When a lightning stroke is forming near a plane, the extremities of the plane send out traveling sparks to meet that lightning and cause it to pass through the plane, much as a lightning rod draws the lightning to itself. Thus a plane may attract lightning. In addition, a plane may fly into an already existing lightning channel and become a part of the channel, or it may initiate a brand new lightning stroke. The process of lightning initiation is similar to that of the lightning initiation caused by tall conducting structures (see Chapter 6).

In only about half of the reported lightning strikes to commercial aircraft are active thunderstorms present (Ref. 4.4). The other half of the reported strikes occurs in precipitation (ice or rain) which is not related to a thunderstorm. Typically, in these cases no lightning is observed before the strike which hits the plane. There appears to be no difference between thunderstorm and non-thunderstorm lightning as regards damaging effects to an aircraft. Undoubtedly, thunderstorm-related lightning strikes would represent a much higher percentage of total strikes if pilots did not intentionally avoid flying through thunderstorms.

It is interesting to note that Apollo 12 (launched for the moon on November 14, 1969) was twice struck by lightning in the first minute after lift-off, once at 6000 ft and again at 13,000 ft. The clouds through which the 360 ft rocket passed had not previously produced any lightning. It is therefore likely that the rocket and its conducting exhaust plume initiated the two discharges. Apollo 12 survived the strikes although their effects were considerable. They included: (1) temporary disconnection of the fuel cells which powered the command module systems, (2) temporary disruption of the primary inertial guidance system used to put the spacecraft into and out of earth orbit and on its way toward the moon, and (3) permanent loss of a number of measurements of the skin temperature of the rocket and of the quantity of propulsion fuel remaining (Ref. 4.6).

From 1964 to 1966 a F-100F jet fighter aircraft, specially protected against the hazardous effects of lightning and equipped to photograph lightning and to measure its electrical current, pressure wave (see Chapter 12), and other properties, was deliberately flown through thunderstorms (Refs. 4.2, 4.7). The plane was struck by lightning 55 times and much valuable information was accumulated.

Ball lightning inside airplanes is discussed in Chapter 15.

REFERENCES

4.1. Conference on Fire Safety Measures for Aircraft Fuel Systems: Report of Conference, December 11-12, 1967, Federal Aviation Administration, Washington, D. C., 1967. Available from Defense Documentation Center as AD 672036.

4.2. Lightning and Static Electricity Conference, 3-5 December 1968, Part II. Conference Papers, May 1969, Technical Report AFAL-TR-68-290, Part II, Air Force Avionics Laboratory, Wright-Patterson Air Force Base, Ohio. Available from Defense Documentation Center as AD 693135.

4.3. *Los Angeles Times,* Final Edition, First Page, January 29, 1969.

4.4. Harrison, Henry T., United Air Line Turbojet Experience with Electrical Discharges, UAL Meteorological Circular No. 57, January 1, 1965.

4.5. Aircraft Accident Report, Boeing 707-121, N709PA Pan American World Airways, Inc., near Elkton, Maryland, December 8, 1963, Civil Aeronautics Board File No. 1-0015, February 25, 1965.

4.6. Godfrey, R., E. R. Mathews, and J. A. McDivitt, Analysis of Apollo 12 Lightning Incident, NASA MSC-01540, January 1970.

4.7. Petterson, B. J., and W. R. Wood, Measurements of Lightning Strikes to Aircraft, Report No. SC-M-67-549, Sandia Laboratory, Albuquerque, New Mexico, January 1968.

ADDITIONAL READING

Fisher, F. A., and J. A. Plumer, *Lightning Protection of Aircraft,* NASA Reference Publication 1008, 550 pages, October 1977. For sale by the National Technical Information Service, Springfield, Virginia 22161.

Clifford, D. W., Aircraft Mishap Experience from Atmospheric Electricity Hazards, Atmospheric Electricity—Aircraft Interaction, 1980, AGARD Lecture Series No. 110, Document No. AGARD-LS-110, for sale by the National Technical Information Service, Springfield, Virginia 22161.

chapter
five

How Does Lightning Damage Trees And
Buildings?

As discussed in Chapter 7, roughly 2000 thunderstorms are in progress over the earth's surface at a given time, and collectively they may produce as many as 100 cloud-to-ground discharges each second — or more than 8 million per day. If these discharges were evenly distributed over the earth, about six percent of them would strike in the world's forested lands (Ref. 5.1). Fortunately, most discharges striking in forests do not cause forest fires. Nevertheless, about 10,000 forest fires are ignited by lightning in the United States each year.

Lightning may strike a tree and leave it apparently unharmed (Ref. 5.2), or it may cause considerable structural damage without noticeable burning. The detailed effects of lightning on trees have been recently studied by Alan R. Taylor of the U.S. Forest Service (Refs. 5.1, 5.3, 5.4, 5.5,

5.6). He found that most trees struck are not killed. The majority recover from whatever lightning damage they have sustained, though many are weakened and ultimately succumb to attacks by insects and disease. Visible damage to tree trunks ranges from superficial bark flaking, to strip-like furrowing along the trunk, to almost total destruction. Damage of intermediate severity is shown in Fig. 5.1: the tree top has been shattered away and a spiral scar with a crack along its axis is wound around the trunk. A close-up of how typical lightning tree scars are formed is shown in Fig. 5.2. Figures 5.3 and 5.4 show lightning-damaged trees.

Taylor, in one study, examined 1000 lightning-damaged Douglas firs in western Montana. Most had shallow continuous scars a few inches wide along their trunks. About 20% had two or more scars, 10% had severed tops, and about 1% had been reduced to slabs and slivers. Most of the scars were spiral, a few were straight. The average scar extended along 80% of the tree height, but none extended to the very tips of the trees. Scars either reached to ground level or close to ground level. Often along the center line of the lightning scar was a crack which penetrated into the tree (Figs. 5.1 and 5.2), and, when wood was removed from a tree by lightning, it was usually ejected as two parallel slabs, separated along this crack. Sometimes, in place of the crack lightning left a narrow strip of shredded inner-bark fiber fixed in a smooth shallow groove about 1/16-in. wide (Fig. 5.5).

These damage characteristics are typical not only for Douglas firs but for most conifers (cone bearing trees, mostly evergreens) throughout the United States. The descriptions also appear valid for most rough-barked species of deciduous trees (trees which shed their leaves each year) such as the oaks. The relatively smooth barks of other deciduous trees — birches, for example — present quite different damage characteristics. The major difference is that the bark is not removed in narrow, uniform strips but is torn off in large, somewhat irregular patches or sheets (Fig. 5.6).

Fig. 5.1. The top of this Douglas fir was shattered away and a spiral lightning scar with a crack along its axis winds around the trunk. (Field sketch by A. R. Taylor)

Sometimes a single lightning discharge can kill a group of trees. In a typical group kill, obvious lightning damage is visible on only one or two trees, often near the center of the dying group. As many as 160 trees have been reported killed this way, but in most cases the groups are probably smaller. It is unclear whether lightning does unseen damage to the roots of trees surrounding the struck tree or whether the aerial parts are affected by the discharge (Ref. 5.7). After reviewing the world-wide literature on the group killing of trees by lightning (Ref. 5.1), Taylor notes that, while very few cases have been reported in the United States, instances have been commonly reported in Europe, Australia, Malaysia, and other parts of the world. A similar phenomenon involving lightning and tree groups is reported to occur frequently in parts of the United States. Entomologists (insect researchers) report that several species of bark beetles attack single trees damaged by lightning and then proceed to

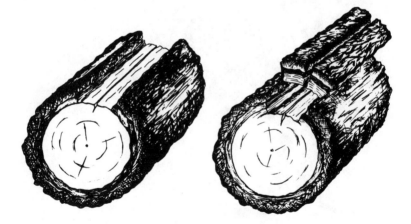

Fig. 5.2. Formation of typical lightning tree scars. Left, bark and wood loss; right, bark loss only. Bark and wood removed in this way are usually found in two slabs. The slabs have been separated along the crack at the centerline of the tree scar. (Sketch by A. R. Taylor)

Fig. 5.3. A Douglas fir with three spiral lightning scars. Only bark was removed, exposing the outer wood. (U.S. Forest Service photo)

attack other trees surrounding the damaged one. Often the result is a group kill similar to those not involving bark beetles — so similar in fact that it has recently been suggested that lightning may be playing a hidden role in this type of group kill; that is, it may do unseen damage to the trees surrounding an obviously struck tree, thereby reducing their natural resistance to attack by the bark beetles and hastening their demise (Refs. 5.6, 5.8).

To understand the direct effects of lightning on trees and buildings, we must first take a closer look at the lightning discharge. In the course of this discussion we will define more specifically some of the words used in the technical literature to describe lightning.

A complete lightning discharge is called a *flash*. A lightning flash typically lasts a few tenths of a second. Each flash between cloud and ground is composed of a number of

Fig. 5.4. Forest Service scientists examine the remains of a 70-ft fir demolished by lightning in western Montana. (U.S. Forest Service photo)

component strokes. The number is sometimes one (a single-stroke flash), most often is three or four (multiple-stroke flash), and may be as many as 20 or 30. Strokes are typically 40 or 50 thousandths of a second apart. Cloud-to-ground lightning often appears to flicker due to the fact that the lightning channel is dark between the component strokes. The longer the time between strokes, the greater the flickering effect. When strokes are very close together (e.g., when they are separated by only 20 thousandths of a second), little flickering is observed because the eye retains the image of the stroke during the "dark time". It is this slow response of the eye that makes movies and TV appear continuous rather than the series of individual pictures that comprise them. Movies that are run too slowly also appear to flicker.

The bright stroke channels we see (an example is shown in Fig. 2.1c) and the stroke currents which cause the light are

Fig. 5.5. Close-up of a lightning tree scar showing a narrow strip of inner-bark fiber along the centerline of a bark-depth furrow. (U.S. Forest Service photo)

Fig. 5.6. Lightning removed bark from this paper birch in large, irregular patches. (U.S. Forest Service photo)

initiated by electrical discharges which move from cloud to ground (a discharge preceding the first stroke of a flash is shown in Fig. 2.1a,b). (Details of the lightning stroke formation are given in Chapters 6 and 9.)

Typical lightning currents measured at ground are shown in Fig. 5.7 (Ref. 5.9). Peak currents are generally 10,000 to 20,000 amps, but occasionally they range up to hundreds of thousands of amps. (The conventional household electric circuit will carry up to 15 amps. A 100 watt light bulb uses about 1 amp.) First strokes generally have larger currents than subsequent strokes. A lightning stroke reaches peak current in a few millionths of a second. The current then decreases, generally terminating in a thousandth of a second or so unless continuing current flows. Figure 5.7a shows a three-stroke flash with no continuing current. Figure 5.7b

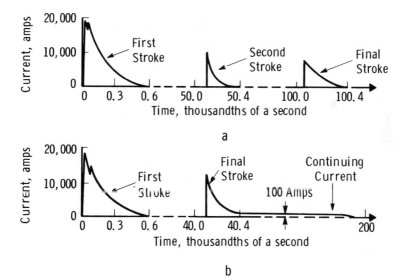

Fig. 5.7. (a) Current at ground due to a typical lightning flash without continuing current — so-called cold lightning. (b) Current at ground due to a typical lightning flash with continuing current — so-called hot lightning.

shows a two-stroke flash in which the final stroke is followed by continuing current. Continuing current is typically 100 amps and may last for one or two tenths of a second. About half of all flashes contain a continuing current following one of the component strokes.

Lightning flashes which contain continuing current are commonly called *hot lightning;* flashes which do not are called *cold lightning.* Hot lightning sets fires, cold lightning does not. The reason for the terminology is obvious. It is, nevertheless, misleading. The temperature in both hot and cold lightning is the same — 15,000 to 60,000°F (see Chapter 11). The difference is that it is maintained longer in hot lightning (tenths of a second vs. thousandths for cold lightning). To set fire to a piece of wood, the flame must contact the wood for a reasonable length of time: initiation of burning depends on the length of time a given temperature is applied. In general, the higher the temperature, the shorter the time. The continuing current from hot lightning provides a high enough temperature for a sufficient time to ignite woody fuels. A U.S. Forest Service study has shown conclusively that continuing currents can cause forest fires (Ref. 5.10).

While so-called cold lightning does not set fires, it can be very destructive. We can think of the bottom end of the lightning channel as a source of current which is forced into the lightning attachment point. This current generates heat in the object through which it flows. The amount of heat depends on the object's resistance to electrical current. If the object has a low resistance (e.g., a metal) (a material with low resistance is called a conductor), there is relatively little heating. If the object has a high resistance (e.g., a piece of wood or plastic), there is a great deal of heating although not necessarily enough to cause burning. The essentially instantaneous lightning current rise inside a high-resistance material causes rapid heating and consequent vaporization (i.e., conversion from solid to gas) of some of the internal material. As a result, a very high pressure is quickly generated within the material and, typically, this pressure blows the material apart.

For many of the tree scars examined by Taylor, the lightning apparently followed a path through the cambium (a thin layer of living cells between the inner bark and the wood) or through the moist inner bark tissue. Possibly these zones were chosen as the lightning's paths because they offered lower resistance than the outer bark or the wood. In any event, the pressure generated in the cambial region blew off a bark strip creating a scar as shown in Fig. 5.5. The pressure also could cause a split in the tree, but it is not clear how a strip of inner-bark fiber could be pasted along the centerline of the scar. In some of the trees examined by Taylor, wood as well as bark was blown out from the tree. In these cases the lightning current apparently traveled deeper within the trunk. Taylor found that old trees (over 200 years) were more likely to suffer wood-loss scars and suggested that perhaps the wood in old trees offers less electrical resistance to current flow than does the cambial zone.

It would seem obvious that lightning-induced forest fires should be suppressed, but this is not necessarily the case (Ref. 5.11). Until recently, frequent fires kept the California forest floor clean; the fires themselves were small and did not damage the trees. Ironically, efforts to prevent and contain forest fires in California enabled the brush to grow more thickly and now most fires are big ones. We may be indebted to ancient forest fires for California's giant sequoias. The seedlings of these trees can germinate in ashes but are suppressed under the thick layer of needles that might cover an unburned forest floor.

From all the foregoing discussion it is obvious that houses can be set on fire by lightning and that high-resistance material comprising houses can be blown apart. Besides directly setting fire to houses and other structures, lightning can indirectly set fires by causing the power company's current to jump through the air between electrical wires or from a wire to a grounded object. The resulting electrical discharge may have a considerably longer duration than the lightning flash which started it. The only known way to protect a house against a direct lightning strike is with a lightning rod system (see Chapter 2). Similarly, the only known way to protect a tree against lightning is to outfit it with a similar rod system. This would consist of metal wire or strip fixed to the tree from a well-established ground and running up the tree trunk to the top of the tree and out the major branches. Thus lightning striking the tree top or any major branch would be harmlessly carried to earth.

REFERENCES

5.1. Taylor, A. R., Lightning Effects on the Forest Complex, *Proc. 9th Annual Tall Timbers Fire Ecol. Conf.*, 127-150 (1969).
5.2. Orville, R. E., Photograph of a Close Lightning Flash, *Science, 162*, 666-667 (1968). The photograph from this article appears on the frontispiece.
5.3. Taylor, A. R., Lightning Damage to Trees in Montana, *Weatherwise, 17* (No. 2), 62-65 (1964).
5.4. Taylor, A. R., Diameter of Lightning as Indicated by Tree Scars, *J. Geophys. Res., 70*, 5693-5695 (1965).
5.5. Taylor, A. R., Tree-bole Ignition in Superimposed Lightning Scars, USDA Forest Serv. Res. Note INT-90, 4 pp. (1969).
5.6. Schmitz, R. F., and A. R. Taylor, An Instance of Lightning Damage and Infestation of Ponderosa Pines by the Pine Engraver Beetle in Montana, USDA Forest Serv. Res. Note INT-88, 8 pp. (1969).
5.7. Minko, G., Lightning in Radiata Pine Stands in Northeastern Victoria, *Australian Forester, 30* (No. 4), 257-267 (1966).
5.8. Komarek, E. V., Sr., The Natural History of Lightning, *Proc. 3rd Annual Tall Timbers Fire Ecol. Conf.*, 139-186 (1964).
5.9. Uman, M. A., *Lightning*, McGraw-Hill Book Co., Inc., New York, 1969, pp. 114-137.
5.10. Fuquay, D. M., R. G. Baughman, A. R. Taylor, and R. G. Hawe, Characteristics of Seven Lightning Discharges That Caused Forest Fires, *J. Geophy. Res., 72*, 6371-6373 (1967).
5.11. Love, R. M., The Rangelands of the Western United States, *Sci. Am., 222*, 88-96, February 1970.

chapter six

Does Lightning "Never Strike Twice?"

Does It Always Strike The Tallest Object?

Much of what is known about lightning today has been discovered precisely because lightning *does* strike the same structure over and over again. To measure a given property of many lightning strokes (for example, the range of values for the peak current), it is necessary to position appropriate instrumentation close to many lightning strokes. The Empire State Building in New York City is struck by lightning an average of about 23 times per year. As many as 48 strikes have been recorded in one year, and during one thunderstorm eight strikes occurred within 24 minutes (Refs. 6.1, 6.2). The General Electric Company conducted a 10-year study (1935-1941, 1947-1949) of lightning at the Empire State Building. Various properties of the lightning current were measured as the current flowed into the building, and high-speed photographs of the formation of the lightning channel near the top of the building were obtained. A similar but more detailed study of lightning was begun on Mount San Salvatore near Lugano, Switzerland in 1942 and still continues at the time of the writing of this book. The instrumented towers near Lugano have recorded lightning current properties an average of about 40 times a year, with a maximum in excess of 100 times a year (Ref. 6.3). A sizable fraction of what we know about lightning currents and, in fact, about lightning in general was determined by the Empire State Building and Lugano studies.

The probability of a structure being struck by lightning depends on its height. For a structure projecting above a moderately flat area in a region of moderate thunderstorm activity, say Pennsylvania or New York (see Chapter 7), the expected number of strikes per year is roughly one for a 300 ft structure, three for a 600 ft structure, five for an 800 ft structure, ten for a 1000 ft structure, and twenty for a 1200 ft structure (Ref. 6.4). A 50 ft structure will be struck about once every four to six years (Refs. 6.2, 6.4), and a quarter acre of flat land (a large residential lot) will be struck about once every 100 years or more (see Chapter 7). Any structure, no matter what its size, may be struck by lightning. Sometimes, however, it is necessary to rebuild the structure before it can be struck again. The bell tower of St. Mark's in Venice was completely destroyed by lightning three times and severely damaged nine times in a period of about 400 years. In 1766 a lightning rod was installed on the tower and since then no further lightning damage has occurred (Refs. 6.5, 6.6).

It is interesting to note that as late as the 18th century church bells were rung in Europe as an effort to protect against lightning. The "ignorant" believed that the ringing of church bells dispersed the evil spirits; the "informed" believed (erroneously) that the sound produced by the bells broke the continuity of the lightning path. Medieval church bells are often inscribed *Fulgura frango* (I break the lightning). Ringing church bells during thunderstorms proved very hazardous to the bell-ringers. In a period of 33 years, 386 church towers were struck, and 103 bell-ringers were killed at their ropes (Refs. 6.5, 6.6).

Since lightning preferentially strikes the taller structures, it is reasonable to ask whether tall structures merely act as lightning rods or whether they influence the lightning to a greater extent. The question is important since much of what we know about lightning comes from studies of lightning strikes to tall buildings. Scientists and engineers would like to be certain that lightning characteristics so obtained also apply to lightning which strikes small buildings, power lines, missiles, etc. The Empire State Building study was the first to show that most of the flashes to a very tall structure are initiated by the structure itself. In the Lugano study about 75% of the flashes observed were initiated by the towers. These structure-initiated flashes differ significantly from the usual cloud-to-ground flash. Flashes to very tall structures which are not structure-initiated are thought to resemble the usual lightning to ground, but there is still some dispute about this.

It is appropriate now to look at how lightning gets started. The usual flash between cloud and ground is initiated in the base of the cloud. The initiating discharge, the downward-traveling spark shown in Fig. 2.1a,b, is called the *stepped leader.* The stepped leader is a low-luminosity traveling spark which moves from the cloud to the ground in rapid steps about 50 yards long. Each step takes less than a millionth of a second, and the time between steps is about 50 millionths of a second. The stepped leader is not seen by the eye, but may be photographed with special cameras (see Chapter 16, Fig. 16.3). The visible lightning flash discussed in the previous chapter occurs when the stepped leader contacts the ground. Precisely what happens when the stepped leader reaches the ground is the subject of Chapter 9. What is important to note now is that the usual stepped leader starts from the cloud without any "knowledge" of what buildings or geography are present below. In fact, it is thought, as suggested in Chapter 2, that the stepped leader is "unaware" of objects beneath it until it is some tens of yards from the eventual strike point. When "awareness" occurs, a traveling spark is initiated from the point to be struck and propagates upward to meet the downward-moving stepped leader, completing the path to ground. In this way the usual lightning flash to ground or to small structures is started. A small fraction of the strikes to very tall buildings also occurs in this way.

However, most lightning flashes to very tall buildings are the result of a reverse process. They are initiated by stepped leaders which start at the building top and propagate upward to the cloud. Because they propagate upward, they and the resulting visible discharge channel branch upward (Fig. 6.1). It is possible that upward-going leaders are just long versions of the leaders (previously called traveling sparks) which propagate upward some tens of yards from lightning rods. That is, the upward-moving stepped leaders from tall buildings may well contact downward-moving leaders beneath or within the cloud. On the other hand, the upward-moving stepped leaders may simply terminate in some region of cloud charge.

Lightning flashes initiated by tall buildings exhibit different current characteristics from those of the usual cloud-to-ground flash discussed in Chapter 5 and illustrated in Fig. 5.7. The current in a structure-initiated flash rises slowly, in hundredths or even tenths of a second, to a current peak of a few hundred amperes and flows for a few tenths of a second. Often the low-level continuous current is punctuated by one or more current peaks similar to those shown in Fig. 5.7 for strokes subsequent to the first. Sometimes the continuous current stops before a current peak occurs. These current peaks are generally larger than those which occur during the continuing current. More details of the characteristics of upward-moving lightning are given in Ref. 6.7.

Fig. 6.1a. Lightning initiated by an upward-moving leader from a tower on Mt. San Salvatore near Lugano, Switzerland. The spot directly beneath the bottom of the lightning channel is a tower light. Upward-initiated lightning is branched upward in contrast to the downward branching of the usual cloud-to-ground lightning flash. (Courtesy, Richard E. Orville, State University of New York at Albany)

Fig. 6.1b. Lightning strikes twice! Another upward-going lightning from the San Salvatore tower. (Courtesy, Richard E. Orville)

It may appear logical from the content of this chapter that lightning should always "hit the tallest object". (Since the point of "strike" is determined by a leader which moves upward from the strike point, the strike point could just as well be called the "point of initiation".) Occasionally it doesn't! The Empire State Building has been struck 50 ft below the top (Ref. 6.2). The usual explanation for such anomalous behavior is the following: It is thought that the greater the electrical charge residing on the downward-moving stepped leader, the longer will be the connecting leader propagating upward from the building top. Thus, a relatively weak stepped leader can come closer to a building top without drawing an upward leader than can a relatively strong stepped leader. It is possible, therefore, that a weak stepped leader might "sneak" past the top of a building and only draw a leader when it had reached some lower level. Another explanation for the anomalous behavior could be that the obvious strike point has been kept from generating an upward-moving leader by a pocket of airborne charge, so-called *space charge.* Properly distributed regions of space charge could cause the lightning to strike almost anywhere. In any event, the cone of protection around the "highest point" (see Chapter 2) almost always describes the zone in which lightning will not strike.

REFERENCES

6.1. Hagenguth, J. H., and J. G. Anderson, Lightning to the Empire State Building — Part 3, *AIEE, 71,* pt. 3, 641-649 (1952).

6.2. Towne, H. M., Lightning, Its Behavior and What to Do About It, 1956, pamphlet published by United Lightning Protection Assn., Inc., Box 9, Onondaga, New York.

6.3. Berger, K., Novel Observations on Lightning Discharges: Results of Research on Mount San Salvatore, *J. Frankl. Inst., 283,* 478-525 (1967).

6.4. McCann, G. D., The Measurement of Lightning Currents in Direct Strokes, *Trans. AIEE, 63,* 1157-1164 (1944).

6.5. Hepburn, F., Man's Study of Lightning, *Science Progress, 44,* 635-646 (1956).

6.6. Schonland, B. F. J., *The Flight of Thunderbolts,* Second Edition, Clarendon Press, Oxford, 1964.

6.7. Uman, M. A., *Lightning,* Dover Publications, Inc., New York, 1984. Sections 2.4.2, 2.5.4, and 4.4.

chapter seven

How Are Thunderstorms Formed?

Are There Locations With No Lightning?

How Many Thunderstorms Are In Progress In The World At One Time?

Roughly 2000 thunderstorms are in progress in the world at any one time (Ref. 7.1). A typical storm is thought to produce one to three cloud-to-ground flashes each minute; and thus, considering all thunderstorms in progress, there are perhaps 30 to 100 flashes to ground every second (Refs. 7.1, 7.2). The statistics on the number of thunderstorms in progress in the world at any one time were obtained in the 1920s by examining the records of meteorological stations located around the world. Where there were no meteorological stations, estimates were made of the thunderstorm activity. For this reason the number 2000 may be somewhat in error. Further, estimates of the number of cloud-to-ground flashes per minute for a typical storm are based primarily on data obtained at relatively few locations throughout the world and may also be in error. For example, very little is known about the lightning or other properties of storms which occur in the tropical regions of South America and South Africa. With the help of orbiting satellites, a more accurate determination of world-wide thunderstorm activity is possible and will probably be made in the near future.

A *thunderstorm-day* for a given location is defined as a day on which thunder is heard at that location, independent of whether one or many thunders were heard. Thunder cannot usually be heard if the lightning causing it is more than about 15 miles away (see Chapter 12). In Fig. 7.1 and 7.2, thunderstorm-day maps are shown for the U.S. and the world, respectively. Each curved line represents a fixed and stated number of thunderstorm-days *per year* and is drawn through all geographical locations that have that particular number of thunderstorm-days. One of the most active regions on earth is Java where thunder is heard 223 days per year. In the U.S., central Florida has the highest number of thunderstorm-days per year with 90. The lowest number occurs along the Pacific coast region of northern California, Oregon, and Washington where thunderstorms and lightning are rare.

A more pertinent statistic than the number of thunderstorm-days per year is the number of lightning strikes per square mile per year. While it is relatively easy to determine thunderstorm days (one needs merely to listen), it is a more difficult task to determine accurately the number of strikes per square mile. The strikes per area have been determined (1) from photographs taken sequentially over periods of several years, (2) from records of strikes to power lines, and (3) from electrical lightning counters which respond to the radio waves emitted by lightning. The combined results of several studies indicate that the number of flashes to ground per square mile per year is equal to between 0.05 and 0.8 times the number of thunderstorm-days per year (Refs. 7.3, 7.4, 7.5). For example, if you live in a region of the U.S. with 40 thunderstorm-days per year, you can expect between 2

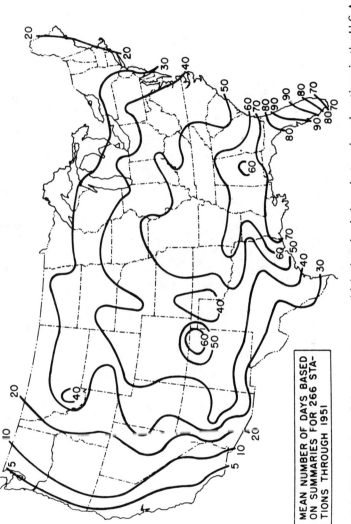

Fig. 7.1. The number of days per year on which thunder is heard at various locations in the U.S.A. Adapted from "Mean Number of Thunderstorm Days in the United States", Technical Paper No. 19, Climatological Services Division, Weather Bureau, September 1952.

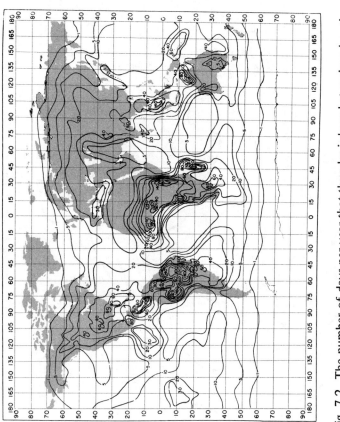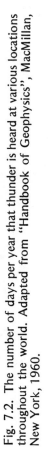

Fig. 7.2. The number of days per year that thunder is heard at various locations throughout the world. Adapted from "Handbook of Geophysics", MacMillan, New York, 1960.

and 30 strikes per year in a square mile area surrounding your home. A lightning strike within a square mile area produces thunder which is heard less than about 3 seconds after the flash (see Chapter 12). You might try to determine for yourself the number of thunderstorm-days and strikes per square mile occurring in your area each year.

As is evident from Figs. 7.1 and 7.2, some parts of the world have a great deal of lightning while others have essentially no lightning. The degree of thunderstorm activity in a particular geographical area depends on its prevailing meteorological conditions.

Thunderstorms can usually be classified as (1) local or convective, or (2) organized traveling. Local thunderstorms form independently of one another and have lifetimes of an hour or two. They produce moderate lightning activity, moderate winds, rain, and sometimes small hail. Most lightning research has been concerned with lightning produced by local thunderstorms because they are relatively easy to study. A local thunderstorm is shown in Fig. 7.3. Organized thunderstorms are violent storms. They may last for many hours producing severe lightning activity, hail over an inch in diameter, winds exceeding 60 mph, and sometimes tornadoes. Such storms are often associated with cold fronts. They may line up along fronts for distances of tens or even hundreds of miles and travel along with the fronts.

What conditions lead to the formation of thunderstorms, and how are those thunderstorms formed? Consider the local or convective thunderstorm. The conditions necessary for the formation of a local thunderstorm are: (1) the air from ground level to many thousands of feet must be moist, (2) the atmosphere must be "unstable" (defined in the next paragraph), and (3) the sun must heat the ground and the air near the ground. In an unstable atmosphere, hot, wet air near the ground will rise to heights where the temperature is below freezing, eventually forming a thundercloud. Let's look now at what is meant by an unstable atmosphere.

If a volume or *parcel* of air is to rise in the atmosphere, it must be lighter than the air which surrounds it. Hot air balloons rise because hot air is less dense (lighter) than normal air. (Helium filled balloons rise because helium is less dense than air.) For a parcel of warm air to continue rising, it must always be at a higher temperature than the surrounding air. Temperature decreases with altitude in the atmosphere, so that the warm air parcel from the ground might be expected to rise continuously. This is not necessarily the case, however, since as the parcel rises it encounters decreasing air pressure. As a result, the parcel expands, and the act of expansion causes the parcel to cool. If the water vapor in the air parcel does not condense, then the cooling rate as the parcel rises will be about 5.5°F per 1000 ft. The air temperature must decrease faster than this if the parcel is to keep moving upward. The air temperature does not often decrease this rapidly much above the earth's surface. Fortunately for the thunderstorm, if the air parcel is moist,

Fig. 7.3. A local or convective thunderstorm at Socorro, New Mexico. (Courtesy, Marx Brook, New Mexico Institute of Mines and Technology)

water vapor will condense as the parcel rises, and the process of condensation releases heat into the parcel. A moist parcel of air will cool at about 3.3°F per 1000 ft of ascent. If the atmospheric temperature decreases faster than 3.3°F per 1000 ft, the atmosphere is said to be "unstable". The parcel will continue to rise.

There are three stages in the life of a local or convective thundercloud: cumulus, mature, and dissipating. The cumulus stage begins when small, fluffy white, cumulus clouds form from rising air parcels. In time, many cumulus clouds combine to form a larger cloud, a cumulus congestus, perhaps a mile in diameter, with a well-defined top which rises at 1000 to 2000 ft per minute. The primary characteristic of the cumulus stage is that air motion throughout the cloud is upward. The cumulus stage lasts 10 to 15 minutes. As the cumulus congestus cloud grows, water drops and ice form within the cloud. Eventually the rain, hail, and snow within the cloud become sufficiently heavy that they can no longer be moved upward by the prevailing updrafts.

The formation of heavy precipitation signals the beginning of the mature stage. Precipitation begins to fall, dragging air downward with it. The cloud, now a cumulonimbus, contains both updrafts and downdrafts, and rain from the cloud reaches the ground. The top of a mature thundercloud may extend to 60,000 ft. It flattens out and assumes a characteristic anvil shape on reaching the stratosphere, the region of the atmosphere in which temperature is constant or increases with height. The mature stage of the storm lasts 15 to 30 minutes and is accompanied by considerable lightning activity. Finally the storm enters its dissipating stage. The violent updrafts and downdrafts decrease, and precipitation is less intense. The water droplets in the cloud evaporate, and the remainder of the cloud is blown away. The dissipating stage lasts about 30 minutes. The total lifetime of the convective thundercloud is roughly an hour.

A single convective thundercloud is technically referred to as a *cell*. Before one thundercloud cell has dissipated others may form. If the various cells form in close proximity, it may be impossible to distinguish between them from the ground. The result is a total storm of relatively long time duration and with relatively constant lightning activity — actually the combined result of the life cycles of many cells. An average local thunderstorm incorporates perhaps three thundercloud cells. Each is electrically active for about 15 to 30 minutes (during its mature stage) and, if the electrically active periods of the three cells do not overlap much, the storm produces appreciable lightning for about an hour and has a total duration of about two hours.

We look now at the traveling organized thunderstorms which produce severe weather. Typically, lines of severe thunderstorms are formed near the boundary of a moving cold front and stationary warm moist air. The cold front, a large mass of cold air, pushes beneath the warm moist air and forces it to rise. The rising air creates thunderstorms along the front. For the forming thunderstorms to become severe, the meteorological conditions must usually differ somewhat from those during the formation of local thunderstorms. Prior to the formation of severe thunderstorms there is often a warm dry layer of air aloft (and a region in which temperature increases with height) which tends to hold down the warm moist air near the earth's surface. The air near the ground becomes progressively warmer and more humid. When the cold front causes the warm air to rise, the ability of the dry air to hold down the wet is destroyed and thunderclouds are violently formed. It should be pointed out that sometimes severe thunderstorms occur under conditions that would normally lead to the generation of only local thunderstorms. There is some argument as to whether severe thunderstorms are composed of many individual cells each similar to a convective thundercloud cell or whether the structure of a severe storm is basically different. It appears now that some severe storms are composed of many individual cells each with a fairly well-defined life cycle while other severe storms do not fit this pattern at all.

One of the primary characteristics of the organized storms is that they can propagate themselves. There are several views of how the propagation process takes place. Common to most theories is the idea that cold air brought to ground by downdrafts through a given thundercloud spreads outward and forces warm moist air adjacent to that thundercloud upward generating a new thundercloud. It appears also that in order for severe storms to propagate effectively over long distances it is usually necessary for the winds aloft to be relatively strong and to increase with height. An excellent non-technical description of what is known about traveling thunderstorms, as well as what is known about local thunderstorms, is given in the paperback book *The Thunderstorm* written by L. J. Battan of the University of Arizona (see References).

REFERENCES

7.1. Brooks, C. E. P., The Distribution of Thunderstorms Over the Globe, *Geophys. Mem., London, 24,* (1925).

7.2. Israël, H., Bemerkung zum Energieumsatz im Gewitter, *Geofis. Pur. Appl., 24,* 3-11 (1953).

7.3. Prentice, S. A., and M. W. Robson, Lightning Intensity Studies in the Darwin Area, *Electrical Engineering Transactions, The Institution of Engineers, Australia, 4,* 217-226 (1968). Contains the results of 9 independent studies.

7.4. Pierce, E. T., The Counting of Lightning Flashes, Special Technical Report 49, Stanford Research Institute, Menlo Park, California, June 1968. Available from Defense Documentation Center as AD 682023.

7.5. Hagenguth, J. H., Photographic Study of Lightning, *AIEE Transactions, 66,* 577-585 (1947).

The following non-technical books deal with thunderstorms: *The Nature of Violent Storms,* L. J. Battan, Doubleday and Co., Inc., Garden City, New York, 1961; *The Thunderstorm,* L. J. Battan, Signet Science Library, New American Library, New York, 1964.

The following technical books deal with thunderstorms: *The Thunderstorm,* H. R. Byers and R. R. Braham, Jr., U.S. Government Printing Office, Washington, D. C., 1949; *The Physics of Clouds,* B. J. Mason, Oxford University Press, London, 1957; *Clouds, Rain and Rainmaking,* B. J. Mason, Cambridge University Press, London, 1962; *Thunderstorms,* C. Mogono, Elsevier, New York, 1980.

chapter eight

Does Cloud Lightning Differ From
Cloud-to-Ground Lightning?   Which Is More Common?

Does Lightning Occur Only In Thunderstorms?

There are two principal types of lightning discharges —
flashes which occur between the thundercloud and the earth
(cloud-to-ground discharges) and flashes within the thundercloud (intracloud discharges). Other types of discharges such
as cloud-to-cloud lightning and cloud-to-air lightning also
occur but not very frequently. It is a widespread misconception that cloud-to-cloud lightning is common.

Although the thundercloud is the most common source of
lightning, it is not the only one. Lightning occurs in
snowstorms, in sandstorms, in non-thunderstorm rain and ice
(see Chapter 4), in the ejected material above erupting
volcanoes (Ref. 8.1; Fig. 8.1), near the fireballs created by
nuclear explosions (Fig. 8.2), and apparently even out of the
clear blue sky (giving rise to the expression "a bolt from the
blue"). For lightning to occur, a region of the atmosphere

must attain an electrical charge sufficiently large to cause electrical breakdown of the air. Some charging almost always occurs when any particulate matter (for example, dust or sand) is subjected to strong winds. (An obvious parallel to the frictional charging discussed in Chapter 1.) Within a typical thundercloud a turmoil of wind, water, and ice exists in the presence of a temperature which decreases with height. Small particles are carried upward by the wind; large particles move downward under the dominant influence of gravity. The various ascending and descending particles exhibit different velocities depending on their size. Particles that move at different velocities collide with one another, and out of these interactions, in ways not yet fully understood, emerge light,

Fig. 8.1. Lightning in the volcano cloud over Surtsey, near Iceland, in December 1963. Volcanic eruptions were observed on November 14, 1963 off the southern coast of Iceland in water about 130 yards deep. Within 10 days an island over half a mile long and about 100 yards above sea level was formed. The island was named Surtsey by the Icelandic Government. (Courtesy, Sigurgeir Jónasson, Icelandair)

positively-charged particles which move upward and heavy, negatively-charged particles which move downward.

The probable distribution of cloud charge for a typical South African thundercloud cell is shown in Fig. 8.3 (Ref. 8.2). The upper part of the thundercloud carries a preponderance of positive (+) charge while the lower part carries a net negative (−) charge. The primary region of positive charge is referred to as the P-region, the primary region of

Fig. 8.2. Five lightning flashes induced by an experimental thermonuclear device exploded on October 31, 1952, at Eniwetok in the Pacific. Photograph is frame number 72 (detonation occurs in frame 1) of a 2000 frame-per-second movie taken from 20 miles away. The five flashes were initiated immediately after detonation probably from instrument bunkers projecting above sea level and propagated upward similar to typical flashes from tall buildings (see Chapter 6). All flashes lasted almost one-tenth of a second and apparently none contained subsequent strokes. The tops of the lightning channels bend towards the fireball. Fleecy cumulus clouds are visible with bases at 2000 ft. The clouds were not lightning producers. (Courtesy, U.S. Atomic Energy Commission)

negative charge the N-region. Recent research indicates that the N-region is not necessarily the vertical pillar of negative charge shown in Fig. 8.3, but rather may be a pillar which leans at an appreciable angle from the vertical (Ref. 8.3). In addition to the main cloud charges (P and N), there may be a small pocket of positive charge at the base of the thundercloud (the p-region). Exact values of the charges in the P-, N-, and p-regions are still a matter of controversy. Malan (Ref. 8.2) suggests P = +40 coulombs, N = -40 coulombs, p = +10 coulombs. (One coulomb is the amount of charge which is moved past a given cross-section of a wire when a current of 1 amp flows for 1 second.) Whatever the exact numbers, the charge in the cloud must be more than the charge which flows in the lightning channel since it is the cloud which provides the charge for the channel. A typical lightning flash to ground generally lowers about 25 coulombs of negative charge from the N-region of the cloud to ground. Occasionally discharges occur from the P-region or the p-region to ground.

The typical cloud-to-ground discharge takes place between the N-region of the cloud and the ground, a distance of roughly two miles. Often the cloud-to-ground lightning path occurs in clear (rain-free) air. The cloud-to-ground lightning flash is, as noted in Chapter 5, composed of a number of discrete strokes, the total time duration of the flash being a few tenths of a second. The typical intracloud discharge takes place between the P- and the N-regions of the cloud, over a distance of a mile or two. The discharge path is through ice and supercooled water (water below 32°F which is not frozen) since most of a typical thundercloud is at a temperature below the freezing level. While the total time duration, charge transfer, and length of an intracloud discharge are similar to that of a cloud-to-ground discharge, the discharge processes differ. This is so because of the different environments in which the two discharges occur and because the cloud-to-ground discharge terminates on a conductor (the earth) while the intracloud discharge does not. Instead of being composed of a number of discrete current and light pulses (strokes), the typical intracloud discharge is fundamentally composed of a single slowly-moving spark or leader which bridges the gap between the N- and P-regions in a few tenths of a second. There is controversy as to whether the leader moves up from the N-region and carries negative charge or moves down from the P-region and carries positive charge. Each may in fact occur on different occasions. A low and continuous luminosity is

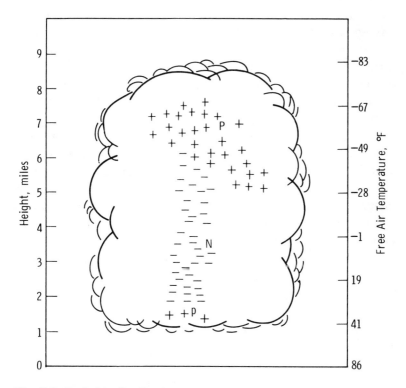

Fig. 8.3. Probable distribution of thundercloud charge according to D. J. Malan (Ref. 8.2).

observed in the cloud during the leader propagation. Superimposed on the continuous luminosity are a number of luminous pulses each lasting about a thousandth of a second. It is thought that these pulses, called K-changes, occur when the propagating leader contacts a pocket of charge opposite in polarity to that of the leader. Much more is known about cloud-to-ground discharges than about intracloud discharges: it is impossible to photograph adequately the intracloud discharge channel since it is obscured by the cloud.

The base of the N-region of a thundercloud is generally situated at the height at which freezing occurs (32°F). The closer the N-region is to ground, the more likely is a cloud-to-ground discharge. Cloud bases are relatively high at the equator where the air near the ground is warm. The height of the cloud base decreases with increasing latitude (north or south) since the air temperature near ground is cooler at higher latitudes. We would expect, therefore, that the number of intracloud discharges relative to the number of cloud-to-ground discharges would be higher at the equator than near the north or south pole. This is indeed the case. Tropical thunderstorms produce roughly 10 intracloud discharges for each cloud-to-ground discharge. In Norway (latitude 60°; see Fig. 7.2) there are about the same number of intracloud discharges as cloud-to-ground discharges. In the United States there are perhaps five intracloud discharges for every cloud-to-ground discharge. All the above numbers are, of course, for typical or average storms. There is a wide variation in the types of lightning produced by individual storms even at a given geographical location.

REFERENCES

8.1. Anderson, R., S. Björnsson, D. C. Blanchard, S. Gathman, J. Hughes, S. Jónasson, C. B. Moore, H. J. Survilas and B. Vonnegut, Electricity in Volcanic Clouds, *Science, 148,* 1179-1189 (1965).

8.2. Malan, D. J., *Physics of Lightning,* The English Universities Press Ltd., London, 1963, pp. 68-75.

8.3. Ogawa, T., and M. Brook, Charge Distribution in Thunderstorm Clouds, *Quart. J. Roy. Met. Soc., 95,* 513-525 (1969).

More details of cloud-to-ground and intracloud lightning are found in: *Lightning,* M. A. Uman, Dover Publications, Inc., New York, 1984.

Additional information on the fraction of strokes to ground as a function of latitude is found in: The Counting of Lightning Flashes, E. T. Pierce, Special Technical Report 49, Stanford Research Institute, Menlo Park, California, June 1968. Available from Defense Documentation Center as AD 682023. Latitudinal Variation of Lightning Parameters, E. T. Pierce, *J. Appl. Meteorology, 9,* 194-195 (1970); The Ratio of Cloud to Cloud-Ground Lightning Flashes in Thunderstorms, S. A. Prentice and D. Mackerras, *J. Appl. Meteor., 16,* 545-555 (1977).

A discussion of the charging processes proposed for thunderclouds is found in: *Atmospheric Electricity,* J. A. Chalmers, 2nd Edition, Pergamon Press, New York, 1967, pp. 399-433; *Thunderstorms,* C. Mogono, Elsevier, New York, 1980; The Electrification of Thunderstorms, J. Latham, *J. Roy. Meteorol. Soc., 107,* 277-298 (1981); Cloud Electrification, R. Lhermitte and E. Williams, *Reviews of Geophysics and Space Science, 21,* 984-992 (1983).

ADDITIONAL READING

Krehbiel, P. R., M. Brook, and R. A. McCrory, An Analysis of the Charge Structure of Lightning Discharges to the Ground, *J. Geophys. Res., 84,* 2432-2456 (1979).

chapter nine

Does A Stroke Between Cloud And Ground Travel Upwards or Downwards?

The answer to the question "Does lightning between cloud and ground go upwards or downwards?" is that, in a sense, it does both. The usual lightning flash between cloud and ground (excluding the discharges initiated by tall structures discussed in Chapter 6) begins with a visually-undetected downward-moving traveling spark called the stepped leader (see Chapters 2 and 6). Since the lightning flash begins with a downward-moving discharge, lightning moves from the cloud to the ground. On the other hand, when the stepped leader reaches ground (or is contacted by an upward-moving discharge some tens of yards above the ground) the leader channel first becomes highly luminous at the ground and then at higher and higher altitudes. The bright, visible channel, or so-called *return stroke,* is formed from the ground up, and one could say, therefore, that visible lightning moves from the ground to the cloud.

It is thought by most lightning researchers that the usual cloud-to-ground discharge begins as a local discharge between the p-charge region in the cloud base and the N-charge region above it (Fig. 9.1, see also Fig. 8.3). This discharge frees *electrons* in the N-region previously immobilized by attachment to water or ice particles. (Electrons are fundamental particles which carry the smallest known unit of negative electrical charge.) Because of their small mass, free electrons are extremely mobile compared to air atoms or to charged ice or water particles. The free electrons overrun the p-region, neutralizing its small positive charge, and then continue their trip toward ground. The vehicle for moving the negative charge to earth is the stepped leader.

Exactly how the stepped leader works is not understood. What is known is that it moves from cloud to ground in rapid luminous steps about 50 yards long. In Figs. 9.1 and 9.2 the luminous steps appear as darkened tips on the less-luminous leader channel which extends upward into the cloud. Each leader step occurs in less than a millionth of a second. The time between steps is about 50 millionths of a second. Negative charge is continuously lowered from the N-region of the cloud into the leader channel. The average velocity of the stepped leader during its trip toward ground is about 75 miles

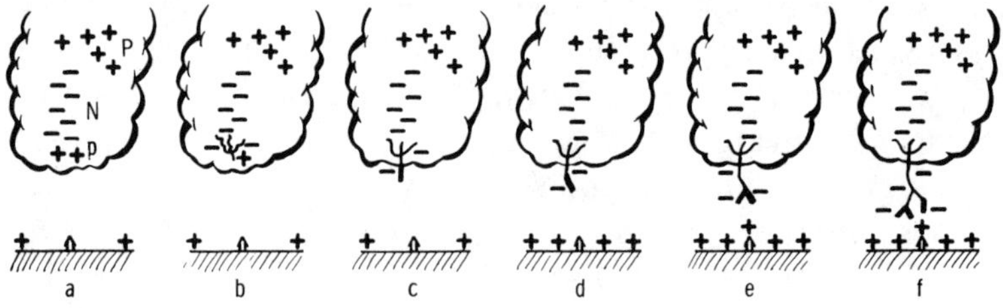

Fig. 9.1. Stepped leader initiation and propagation. (a) Cloud charge distribution just prior to p-N discharge. (b) p-N discharge. (c)-(f) Stepped leader moving toward ground in 50-yard steps. Time between steps is about 50 millionths of a second. Scale of drawing is distorted for illustrative purposes.

per second with the result that the trip between cloud and ground takes about 20 thousandths of a second. A typical stepped leader has about 5 coulombs of negative charge distributed over its length when it is near ground. To establish this charge on the leader channel an average current of about 100 or 200 amperes must flow during the whole leader process. The pulsed currents which flow at the time of the leader steps probably have a peak current of about 1000 amperes. The luminous diameter of the stepped leader has been measured photographically to be between 1 and 10 yards. It is thought that most of the stepped-leader current flows down a narrow conducting core less than an inch in diameter at the center of the observed leader. The large photographed diameter is probably due to a luminous electrical corona surrounding the conducting core.

When the stepped leader is near ground, its relatively large negative charge induces large amounts of positive charge on the earth beneath it and especially on objects projecting above the earth's surface (Fig. 9.2). Since opposite charges attract each other, the large positive charge attempts to join the large negative charge, and in doing so initiates upward-going discharges (Figs. 2.1 and 9.2). One of these upward-

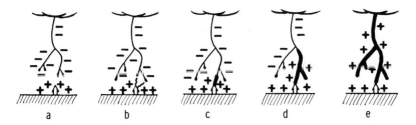

Fig. 9.2. Return stroke initiation and propagation. (a) Final stages of stepped leader descent. (b) Initiation of upward-moving discharges to meet downward-moving leader. (c)-(e) Return stroke propagation from ground to cloud. Return stroke propagation time is about 100 millionths of a second; propagation is continuous. Scale of drawing is distorted.

going discharges contacts the downward-moving leader and thereby determines the lightning strike-point. When the leader is attached to ground, negative charges at the bottom of the channel move violently to ground, causing large currents to flow at ground and causing the channel near ground to become very luminous. Since electrical signals (or any signals, for that matter) have a maximum speed of 186,000 miles per second — the speed of light — the leader channel above ground has no way of knowing for a short time that the leader bottom has touched ground and has become highly luminous. The channel luminosity, the return stroke, propagates continuously up the channel and out the channel branches at a velocity somewhere between 20,000 and 60,000 miles per second, as shown in Fig. 9.2. The trip between ground and cloud takes about 100 millionths of a second. As the return stroke luminosity moves upward so also does the region of high current. When the leader initially touches ground, copious numbers of electrons flow to ground from the channel base. As the return stroke moves upward, large numbers of electrons flow at greater and greater heights. Electrons at all points in the channel *always move downward* even though the region of high current and high luminosity moves upward. Eventually, in some thousandths of a second, the coulombs of charge which were on the leader channel have all flowed into the ground. The current at ground associated with this charge transfer is shown in Fig. 5.7. Since the return stroke channel is a good conductor and is tied to the ground, it will become positively charged like the ground in response to the negative charge in the lower part of the cloud.

It is the return stroke that produces the bright channel of high temperature (Chapter 11) that we see. The eye is not fast enough to resolve the propagation of the return stroke and it seems as if all points on the channel become bright simultaneously. The reason we do not visually detect the stepped leader preceding a first return stroke is because the eye cannot resolve the time between when the weakly luminous leader is formed and the bright return stroke illuminates the leader channel. The return stroke produces most of the thunder we hear (see Chapter 12).

After the stroke current has ceased to flow, the lightning flash may be ended, in which case the discharge is called a single-stroke flash. As noted in Chapter 5, most flashes contain three or four strokes, typically separated by gaps of 40 or 50 thousandths of a second. Strokes subsequent to the first are initiated only if additional charge is made available to the top of the previous stroke channel less than about 100 thousandths of a second after current has stopped flowing in the previous stroke. Additional charge can be made available to the channel top by the action of electrical discharges (so called *K-streamers* and *J-streamers*) which move upward from the top of the previous return stroke into higher areas of the N-charge region of the cloud (Fig. 9.3). When this additional charge is available, a continuous (as opposed to stepped) leader, known as a *dart leader*, moves down the defunct return stroke channel again depositing negative charge from the N-region along the channel length. The dart leader thus sets the stage for the second (or any subsequent) return stroke. The dart leader's earthward trip (Fig. 9.4) takes a few thousandths of a second. Because it occurs so close in time to the return stroke, it is not seen by the eye. To special cameras it appears as a luminous section of channel about 50 yards long which travels smoothly earthward at about 1000 miles per second. The dart leader generally deposits somewhat less charge along its path than does the stepped leader, with the result that subsequent return strokes generally lower less charge to ground and have smaller currents (Fig. 5.7) than first strokes.

The first stroke in a flash is usually strongly branched downward because the stepped leader is strongly branched

(Figs. 9.1 and 9.2). Dart leaders generally follow only the main channel of the previous stroke and hence subsequent strokes show little branching (Fig. 9.4).

The time between strokes which follow the same path can be tenths of a second if a continuing current (Chapter 5) flows in the channel between strokes. Apparently, the channel is ripe for a dart leader only after the continuing current has terminated.

A typical cloud-to-ground discharge lowers about 25 coulombs of negative charge from the N-region of the cloud to the earth. This charge is transferred in a few tenths of a second by the three or four component strokes and any continuing current which may flow. While the leader-return stroke process transfers charge to ground in two steps (charge is put on the leader channel and then is discharged to ground), the continuing current represents a relatively steady charge flow between the N-region and ground.

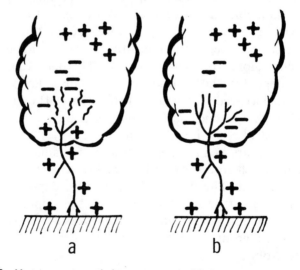

Fig. 9.3. K-streamers and J-streamers making more negative charge available to the channel top during the 50 thousandths of a second or so following the cessation of current flow in the first return stroke. Scale of drawing is distorted.

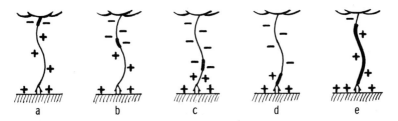

Fig. 9.4. Dart leader and subsequent return stroke. (a)-(c) Dart leader deposits negative charge on defunct first-stroke channel during its thousandth-of-a-second trip to ground. (d)-(e) Subsequent return stroke propagates from ground to cloud in about 100 millionths of a second. Scale of drawing is distorted.

Thus far in this chapter we have discussed the usual stepped leader which lowers negative charge between the cloud and the earth. Occasionally downward-moving stepped leaders are observed that lower positive charge. Currents due to the resulting "positive" return strokes have been measured directly during discharges to instrumented towers. Positive strokes are characterized by rates-of-rise of current at the ground roughly five times slower than those for typical negative strokes, and by charge transfers roughly three times greater than those of typical negative strokes. The maximum measured charge transfer due to a positive stroke is 300 coulombs. Positive strokes are probably initiated between the P-charge region of the cloud and ground when strong winds blow the cloud such that the P-region is brought relatively close to a mountain side or to the earth's surface. Positive discharges rarely consist of more than one stroke.

REFERENCES

The material contained in Chapter 9 is considered in more detail in the following three references presented in order of decreasing technical content: *Lightning,* M. A. Uman, Dover Publications, Inc., New York, 1984. *Physics of Lightning,* D. J. Malan, The English Universities Press Ltd., London, 1963. *The Flight of Thunderbolts,* B. F. J. Schonland, 2nd Edition, Clarendon Press, Oxford, 1964.

chapter ten

How Long And How Wide Is The Lightning Channel?

Why Is Lightning Zig-Zag?

How Can I Best Photograph Lightning?

Let's begin with "how long". Channels longer than 90 miles have been observed (Ref. 10.1). On the other hand, sparks of a few yards length have been seen in and around clouds, and probably even shorter "lightning" exists.

There are basically four ways of determining channel length. (1) The most straightforward way is by photographing lightning and then determining its length from the photograph. Since channels inside clouds cannot be photographed, the value of this technique for determining channel length is somewhat restricted. (2) Radar has been used to measure channel length. A radar set sends out electromagnetic pulses which are reflected back by certain objects (e.g., metallic airplanes). Since the pulses travel at 186,000 miles per second, measuring the elapsed time between the emission of a pulse and the reception of its reflection enables the distance to the reflector to be measured. An actual radar set has a TV-like screen on which appears images of the reflecting objects. A lightning channel is a good conductor and hence reflects radar pulses much as an airplane does. Radar wavelengths can be chosen to minimize reflection of radar pulses by the ice and water in clouds so that a good image of the lightning channel can be obtained on the radar screen. (3) Channel lengths can be determined from the duration of thunder. The minimum possible channel length associated with a thunder duration of T seconds is roughly T/5 miles (see Chapter 12). For example, if thunder lasts 20 seconds, the channel producing it has a length of about 4 miles or more. (4) Finally, there are a number of ways of determining lightning channel height within a cloud from electrical measurements made on the ground (Ref. 10.2). Charge motion in the lightning channel or in the cloud will induce voltages and currents in ground-based instruments enabling properties of the channel and cloud charge to be determined. For example, the vertical height of the J-streamers and K-streamers occurring in the cloud between strokes (Fig. 9.3) can be determined from electrical measurements made at ground.

Now that we know how channel lengths are measured, let's look at the results of the measurements. The average vertical stroke height is about 3 to 4 miles. The maximum vertical extent of the usual stroke to ground is the top of the N-charge region, a height of 6 miles or more. Each stroke in a multiple-stroke flash averages about a third of a mile longer than the preceding one. This is the case because, in order to obtain negative charge for a new stroke, J-streamers and K-streamers tap new areas of the N-charge region of the cloud (Figs. 8.3 and 9.3) during the time between strokes, and these tapped regions become part of the new stroke. The horizontal component of the lightning channel within a cloud or through several clouds may be larger than the vertical component. Horizontal sections of 5 to 10 miles within clouds are not uncommon. Radar studies have shown horizontal channels longer than 90 miles (Ref. 10.1). Long horizontal channels must feed on the charge of many thundercloud cells. In addition to cloud-to-ground discharges with long horizontal in-cloud components or similar long intracloud or cloud-to-cloud discharges which do not reach ground, long lightning discharges have occasionally been observed to occur from the tops of thunderclouds upward (Ref. 10.2). These vertical air discharges may be tens of miles in length.

When we speak of the length of lightning, we normally picture long sections of the channel as being fairly straight. Photographs show, however, that the channel is very tortuous on almost any scale. There are zigs and zags 100 yards long and, within these, other zigs and zags 10 yards long, and within these yet smaller zigs and zags (Refs. 10.3, 10.4, 10.5). Very detailed photographs have shown lightning channels that twist and bend on a distance scale measured in inches (Ref. 10.5). Figure 10.1 shows a section of a lightning channel with a height of about 1000 yards; Fig. 10.2 shows about 200 yards of channel; Fig. 10.3 about 10 yards. If we could grab a long section of the channel and pull on both ends to straighten it out, the channel section might well be two or more times longer than we initially thought.

Lightning diameter measurements have been made in two ways: (1) by examining the interaction between lightning and objects and (2) from photographs. Photographic measurements of lightning almost always overestimate the luminous diameter of the channel: the bright channel overexposes the film making the recorded image broader than it should be. It is extremely difficult to expose a lightning channel photograph properly. The best channel diameter photographs (Ref. 10.5) yield diameters of between about 2 and 7 in., and these values are probably overestimates. Lightning photography is discussed again at the end of this chapter.

Fig. 10.1. About 1000 vertical yards of lightning channel near Tucson, Arizona (Ref. 10.4). Note the large scale channel tortuosity. (Courtesy, H. B. Garrett and A. A. Few, Rice University)

When lightning hits an object such as a tree or a rock, it leaves visible damage. In many cases the damage can be related to channel diameter. Taylor (Ref. 10.6; see also Chapter 5) found that the lightning-caused furrows that spiral along tree trunks are between about 0.6 and 5.0 in. wide. The lightning diameter, therefore, must be roughly this size or smaller. If it were much bigger, it is hard to see how it could create the furrows.

When lightning strikes sand or certain kinds of rocks, the channel heat melts the material along its path. When the melted material solidifies, the resulting *fulgurite* (from the Latin *fulgur,* meaning lightning) represents a permanent record of the lightning diameter and path. Fulgurites in dry

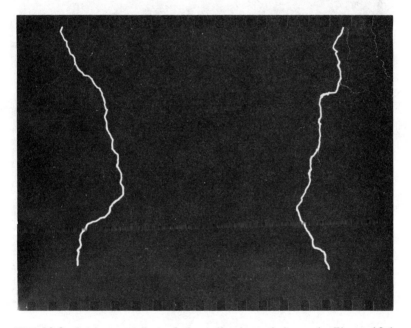

Fig. 10.2. Bottom portion of the split channel shown in Figure 10.1 (Ref. 10.4). Vertical height shown is about 200 yards. Tortuosity on a scale smaller than 10 yards is evident. (Courtesy, Leon E. Salanave, University of Arizona)

sand are generally long hollow tubes with corrugated glassy walls. An artificial fulgurite produced by an electrical discharge in the laboratory is shown in Fig. 10.4. Fulgurites have been traced downwards into sand as much as 20 yards. Diameters are usually 0.5 to 2 in. From the corrugated appearance of fulgurite walls, it is probable that the initial diameter collapses somewhat on cooling. "Fossil" fulgurites 250 million years old have been discovered (Ref. 10.7). The

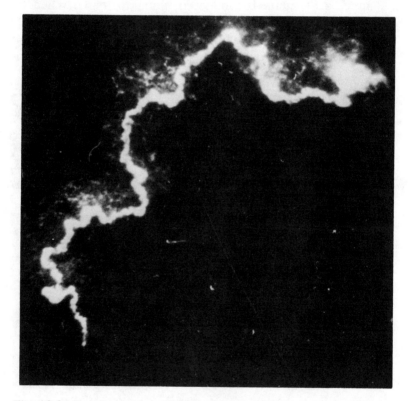

Fig. 10.3. One frame from a high-speed movie of lightning striking a TV tower near Tucson, Arizona (Ref. 10.5). About 10 yards of channel are shown and tortuosity on an inch scale is evident. (Courtesy, W. H. Evans, University of Arizona, and R. L. Walker, University of Florida, Cape Canaveral)

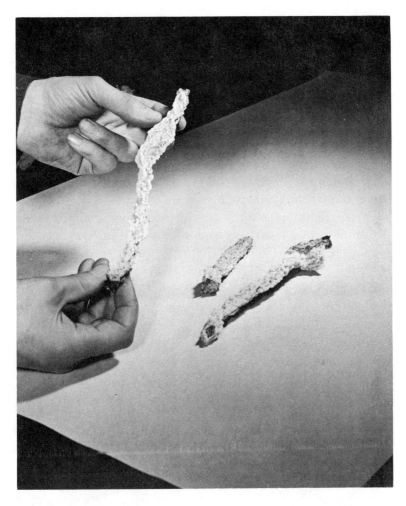

Fig. 10.4. Artificial fulgurites produced by P. L. Bellaschi, Westing-house Electric Corporation, by allowing an electrical discharge to pass through quartz sand. The hollow tubes of solidified sand are close replicas of natural fulgurites.

fulgurite diameter probably does not differ much from the lightning diameter.

The author has measured lightning diameters by allowing lightning to pass through Fiberglas screens on the way to lightning rods and then measuring the size of the holes melted in the screen by the lightning (Ref. 10.8). The screens covered rods on the tops of towers near Tucson, Arizona and near Lugano, Switzerland. About 50 holes were obtained. Figure 10.5 shows typical lightning holes. Diameters were usually about an inch or smaller and sometimes were as small as a sixteenth of an inch. Hole diameters represent the width over which the lightning channel is hot enough for sufficient time to melt or vaporize the screen.

When lightning strikes an electrical conductor such as a lightning rod or an airplane wing, it often leaves a spot

Fig. 10.5. Holes melted in two Fiberglas screens by lightning. At least four strokes passed through the screen on the left. One stroke passed through the screen on the right.

(sometimes a raised lump, sometimes a depression) about an eighth of an inch wide. Examples of such spots on three copper disks are shown in Fig. 10.6. The size of a lightning spot is roughly the diameter of the lightning channel when it enters the conductor. At the contact point the lightning current must flow through a mixture of metal vapor and air, with the result that its diameter is smaller than in the metal-free air away from the conductor. Thus, spot sizes on conductors provide a lower diameter limit. It should be noted that a large lightning current flowing for a relatively long time may do considerably more damage to a conductor than just a spot. A hole may be burned (Figs. 4.1-4.3) or the conductor may be totally melted or vaporized.

Based on all the available evidence, the lightning channel diameter is of the order of an inch. However, for two reasons there is some ambiguity concerning what is meant by lightning diameter. First, there is no definite boundary between the stroke and the air surrounding the stroke; second, the diameter must change with time in response to the changing current.

Fig. 10.6. Spots produced by lightning strikes to three copper disks which were placed on top of TV towers near Tucson, Arizona. The center disk was apparently hit by three lightning strokes, one of which raised a one-eighth inch lump. The disk on the right exhibits a copper lump apparently formed as the lightning channel moved back and forth across the disk.

Why is the lightning channel so tortuous? The answer is not known, but some reasonable guesses may be made. The larger-scale tortuosity in the channel (representing, say, tens of yards or more) is due to the fact that the stepped leader makes such an errant trip to ground. Why does it do this? Possibly various airborne regions of charge (space charge, see Chapter 6) divert the leader on its trip. More likely, the leader just doesn't know exactly where it wants to go, except that ultimately it wants to move downward. The smaller-scale tortuosity in the channel may be formed when the leader steps are formed, or may be formed by action of the magnetic forces associated with the return stroke current. As is probably becoming more and more apparent with each succeeding chapter in this book, there is an awful lot we still don't know about lightning.

Since we have discussed in this chapter the use of photography as a scientific tool to determine lightning channel parameters, it is appropriate to say a few words about how to take lightning photographs. There are basically two ways to photograph lightning using ordinary cameras with no special attachments: (1) by time exposure until the lightning occurs and is over, and (2) by snapping the camera shutter (set at 1/30 sec or slower) when the lightning is first seen.

(1) The camera is put on a tripod or other firm base, pointed in the direction of the storm, and the shutter is left open (time exposure) until lightning occurs in the field of view. Of course, it must be nighttime, otherwise the film will be immediately fogged. The length of time the shutter can remain open without appreciable fogging of the film depends on the ambient light level. Away from city lights with a film speed of ASA-100 (roughly the speed of Kodak Plus-X black and white film or Kodacolor-X color film) and an aperture opening of f/11, several minutes are possible. Camera focus should be set at infinity. Since there are several lightning strokes to ground during each minute of the electrically

active period of a storm (Chapter 7), it is not difficult to obtain photographs. (2) If the lightning flash to be photographed consists of only one stroke, your reflexes will not be fast enough to fire the shutter in time to record the flash. However, if it is a multiple-stroke flash, as most flashes are, you will record a subsequent stroke or strokes. Since subsequent strokes occur at intervals of roughly 1/30 of a second, a shutter speed of 1/30 second or slower is necessary to avoid taking a picture of what goes on between strokes. Since subsequent strokes are generally unbranched (Chapter 9), the photographed lightning will be unbranched and will look somewhat peculiar. Again, for a film speed of ASA-100 try an aperture of f/11 with the focus set at infinity.

The two basic techniques for photographing lightning can be improved by using special equipment with ordinary cameras. (1) Time exposures of lightning in daylight are possible if the proper filter is used (Refs. 10.9, 10.10). Light from the sun is weak at a wavelength of 6563Å* (in the red) because hydrogen in the sun's outer layers absorbs that wavelength, the so-called H_α line. Hydrogen is present in the lightning channel due to the high-temperature break-up of water vapor (H_2O). Since hydrogen in the lightning channel radiates at 6563 Å (H_α) (see Chapter 11), it follows that an H_α filter that passes only light near this wavelength both allows the channel to be photographed and prevents rapid film fogging by daylight. For maximum effectiveness, the film used should have good red sensitivity. Kodak Linagraph Shellburst is perhaps the best for the purpose. In cloudy weather at f/11, a daytime exposure of several minutes should be possible without excessive film fogging. A 2x2 in. H_α filter with a bandwidth of about 40 Å costs about $200. (2) Triggering the camera shutter can be accomplished with instruments that respond to the electrical signals generated in

---

*One Å (Ångstrom) equals one ten-billionth of a centimeter. There are 2.54 centimeters to the inch.

the cloud or by the stepped leader prior to the first return stroke (Refs. 10.11, 10.12). In this way the first stroke can be photographed. Essentially, the shutter is tripped by a solenoid coupled to a radio antenna and appropriate electronics.

REFERENCES

10.1. Ligda, M., The Radar Observation of Lightning, *J. Atmospheric Terrest. Phys., 9*, 329-346 (1956).

10.2. Uman, M. A., *Lightning*, McGraw-Hill Book Company, New York, New York, 1969, pp. 2, 81-85.

10.3. Hill, R. D., Analysis of Irregular Paths of Lightning Channels, *J. Geophys. Res., 73*, 1897-1906 (1968).

10.4. Few, A. A., H. B. Garrett, M. A. Uman, and L. E. Salanave, Comments on Letter by W. W. Troutman 'Numerical Calculation of the Pressure Pulse from a Lightning Stroke', *J. Geophys. Res., 75*, 4192-4195 (1970).

10.5. Evans, W. H., and R. L. Walker, High Speed Photographs of Lightning at Close Range, *J. Geophys. Res., 68*, 4455-4461 (1963).

10.6. Taylor, A. R., Diameter of Lightning as Indicated by Tree Scars, *J. Geophys. Res., 70*, 5693-5695 (1965).

10.7. Harland, W. B., and J. L. F. Hacker, 'Fossil' Lightning Strikes 250 Million Years Ago, *Advancement of Science,22*, 663-671 (1966).

10.8. Uman, M. A., The Diameter of Lightning, *J. Geophys. Res., 69*, 583-585 (1964).

10.9. Salanave, L. E., and M. Brook, Lightning Photography and Counting in Daylight, Using H_α Emission, *J. Geophys. Res., 70*, 1285-1289 (1965).

10.10. Krider, E. P., Comment on Paper by Leon E. Salanave and Marx Brook, 'Lightning Photography and Counting in Daylight, Using H_α Emission,' *J. Geophys. Res., 71*, 675 (1966).

10.11. Schonland, B. F. J., and J. S. Elder, Anticipatory Triggering Devices for Lightning and Static Investigations, *J. Franklin Inst., 231*, 39-47 (1941).

10.12. Hawe, R. G., Electrostatic Trigger Used for Daylight Lightning Photography, *Photographic Science and Engineering, 12*, 219-221 (1968).

chapter eleven

How Hot Is Lightning?

The lightning return stroke is more than four times hotter than the surface of the sun. The peak temperature of lightning is greater than 50,000°F. The surface of the sun has a temperature of about 11,000°F.

Exactly how hot the lightning return stroke channel gets is not known. The best peak temperature measurements made to date represent the average temperature existing during a time interval of a few millionths of a second. For time varying systems, average values are necessarily less than peak values. For example, the lightning temperature could be very high for a few billionths of a second and relatively low for the remainder of the millionths-of-a-second measuring time interval. The resultant average temperature would then be close to the low value. Lightning temperature has been determined by examining the characteristics of the light emitted by the channel.

The lightning channel is composed of very hot air. Air at room temperature is made up primarily of nitrogen and oxygen molecules. A nitrogen molecule consists of two joined nitrogen atoms; an oxygen molecule of two joined oxygen atoms. The nitrogen atom has a core or nucleus containing seven positively-charged particles (protons) and seven uncharged particles (neutrons). Orbiting the nucleus at relatively large radii are seven negatively-charged electrons. (The electrons circle the nucleus much as the earth and other planets circle the sun.) The oxygen atom consists of eight protons, eight neutrons, and eight electrons. As the temperature of air at atmospheric pressure is increased above room temperature to about 12,000°F, the heat causes the molecules to break apart into atoms. With a further increase in temperature to about 30,000°F each atom loses one of its electrons, and with higher temperatures still (say above 60,000°F), more electrons are lost. An atom which has lost an electron or electrons is called an *ion*. If one electron is lost, the atom is said to be singly ionized; if two, doubly ionized, and so on. The lightning channel is composed primarily of nitrogen atoms and ions, oxygen atoms and ions, and free electrons. How do we know this?

If sunlight is passed through a prism, the light is decomposed into its component colors or spectrum. (The rainbow is the sun's spectrum produced by water droplets.) Sunlight is composed of a range of colors varying in hue continuously from red to orange to yellow to green to blue to violet. Incandescent solids (e.g., tungsten filaments) or incandescent gases at very high pressure or occupying large volumes (e.g., the sun) produce continuous spectra; that is, light which includes all colors. Luminous gases at lower pressure or occupying small volumes (e.g., the lightning channel) produce spectra with "lines" at characteristic wavelengths. A given molecule, atom, or ion is capable of emitting radiation only at specific wavelengths which are different for different kinds of particles. Each molecule, atom, or ion has its own unique radiation "signature" by which it can be positively identified. For example, in the visible wavelength region a nitrogen atom radiates primarily one red spectral line while a nitrogen ion radiates primarily a number of blue lines. We can get a rough idea of the temperature of air by determining from the radiation emitted by the air whether it is predominantly composed of molecules, atoms, or ions. A more exact determination can be made by examining the intensities of the characteristic spectral lines from a given molecule, atom, or ion.

A spectrometer is an instrument that disperses the light from a source so that characteristics of the component spectrum can be measured. The dispersion is effected by either a prism or a diffraction grating. A drawing of a lightning spectrometer is shown in Fig. 11.1. Light from the lightning channel passes through the transmission diffraction grating and is split into its spectral components. These components as well as an undiffracted image of the lightning channel are focused by a lens onto the channel isolator slit assembly. For illustrative purposes, the lightning in Fig. 11.1 is assumed to contain only three spectral lines: a red, a green, and a blue. Note that each of the spectral lines has the shape of the lightning channel from which it came. If a stationary piece of photographic film were put in front of the channel isolator slit assembly, a photograph like that in Fig. 11.2 would be obtained. The photograph shown was taken by placing an inexpensive transmission grating† over the lens of a 35 mm camera while taking a time exposure (see Chapter 10) until lightning occurred. In Fig. 11.2, the red line to the far left (the H_α line discussed in the previous chapter) is emitted by hydrogen atoms present in the channel due to the break-up of water vapor (H_2O). Some water vapor is always

---

†A page-size sheet of diffraction grating can be purchased for a few dollars. One supplier is Edmund Scientific Co., Barrington, N.J.

present in the air; quite a lot may be present during a thunderstorm. The red line just to the right of the hydrogen line is emitted by nitrogen atoms. Most of the remaining lines are emitted by singly ionized nitrogen atoms.

Spectra such as that in Fig. 11.2 yield no information about what is happening in the lightning channel as a function of time, since all the light of a given wavelength is recorded at the same place on the film no matter when it is emitted by the channel. With the help of Fig. 11.1 we see how to obtain time-resolved lightning spectra. The channel isolator slit allows the spectrum emitted by a short section of the lightning channel to fall on a piece of moving film at the back of the slit. Thus the spectrum is streaked out in time on the moving film. In the example shown, the green line appears at the earliest time followed by the blue and then the red lines. A time-resolved spectrum of a long spark in air is shown in Fig. 11.3 (Ref. 11.1). All existing time-resolved lightning spectra have been taken on black and white film since it permits much easier intensity measurements. Obtaining lightning spectra is too difficult a job to waste a spectrum by taking it on color film. The spark and lightning spectra differ primarily in that the lightning spectrum is stretched out three to five times longer than the spark spectrum (Ref. 11.1). For example, a nitrogen ion spectral line might last 5 millionths of a second in the spark, whereas for a typical lightning stroke it might last 20 millionths of a second.

From measurements of the intensities of the various spectral lines and use of appropriate theory, channel temperature can be determined as a function of time (Refs. 11.2, 11.3). A graph of typical lightning return stroke temperature vs. time is shown in Fig. 11.4. The peak temperature measured is about 55,000°F. The actual peak temperature could be higher as was previously noted. The temperature falls to about half of peak value in about 30 millionths of a

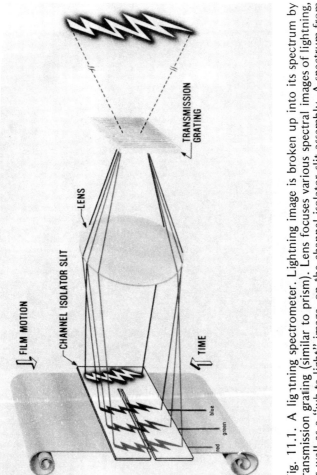

Fig. 11.1. A lightning spectrometer. Lightning image is broken up into its spectrum by transmission grating (similar to prism). Lens focuses various spectral images of lightning, as well as a "wh te light" image, on the channel isolator slit assembly. A spectrum from one section of the channel passes through the slit and is recorded on the moving film.

Fig. 11.2. Lightning spectrum as might be obtained by putting stationary photographic film in front of the channel isolator slit assembly of Fig. 11.1. Actually, the photograph was taken by placing an inexpensive diffraction grating in front of the lens of a 35 mm camera. (Courtesy, Richard E. Orville, State University of New York at Albany) (Reproduced in color on inside front cover)

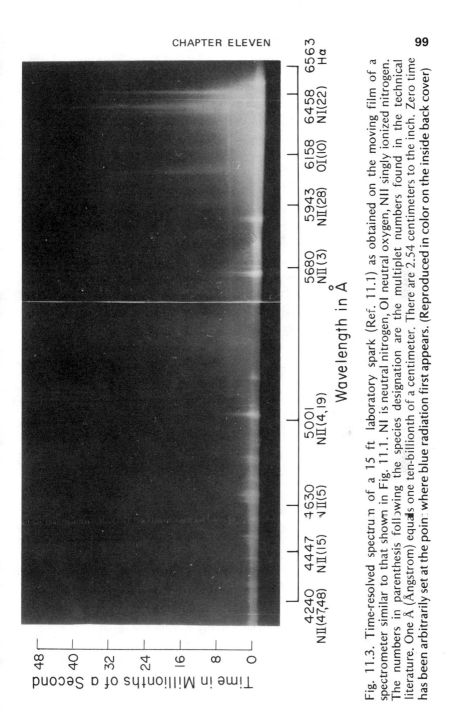

Fig. 11.3. Time-resolved spectrum of a 15 ft laboratory spark (Ref. 11.1) as obtained on the moving film of a spectrometer similar to that shown in Fig. 11.1. NI is neutral nitrogen, OI neutral oxygen, NII singly ionized nitrogen. The numbers in parenthesis following the species designation are the multiplet numbers found in the technical literature. One Å (Ångstrom) equals one ten-billionth of a centimeter. There are 2.54 centimeters to the inch. Zero time has been arbitrarily set at the point where blue radiation first appears. (Reproduced in color on the inside back cover)

second. For a current of about 100 amperes in the lightning channel (associated with the return stroke current tail or with a continuing current — Fig. 5.7) the temperature will be roughly 15,000 to 20,000°F. When the stroke current stops, the channel cools to a few thousand degrees in hundredths of a second (Ref. 11.4).

The temperature of one stepped-leader has been measured spectroscopically (Ref. 11.5). The peak temperature associated with the leader step was about 50,000°F. Temperature in the channel behind the step decreased with distance up the channel and was roughly 25,000°F five step lengths beyond the step.

In addition to allowing a determination of the channel temperature and of the types of particles in the channel, quantitative spectroscopy allows the number densities of all the channel particles to be measured. For example, it has been deter-

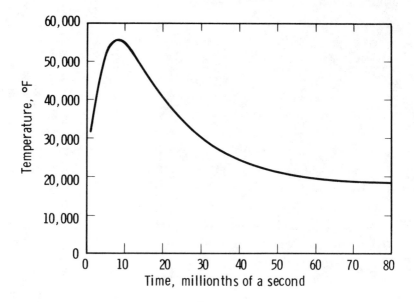

Fig. 11.4. Typical lightning return-stroke temperature as a function of time. (Ref. 11.2).

mined that there are over 1,000,000,000,000,000,000 free electrons per cubic inch in the lightning channel (Refs. 11.2, 11.3).

Also, spectroscopy permits the pressure in the lightning channel to be determined as a function of time. It is the initial high channel pressure (possibly more than 100 times atmospheric pressure) caused by the initial high channel temperature which produces the thunder we hear. We will consider channel pressure and the resultant thunder generation in the next chapter.

REFERENCES

11.1.   Uman, M. A., Comparison of Lightning and a Long Laboratory Spark, *Proc. IEEE, 59,* 457-466 (1971).

11.2.   Uman, M. A., *Lightning,* Dover Publications, Inc., New York, 1984, Chapter 5.

11.3.   Orville, R. E., A High-Speed Time-Resolved Spectroscopic Study of the Lightning Return Stroke, *J. Atmosph. Sci., 25,* 827-856 (1968).

11.4.   Uman, M. A., and R. E. Voshall, Time Interval Between Lightning Strokes and the Initiation of Dart Leaders, *J. Geophys. Res., 73,* 497-506 (1968).

11.5.   Orville, R. E., Spectrum of the Lightning Stepped Leader, *J. Geophys. Res., 73,* 6999-7008 (1968).

ADDITIONAL READING

Orville, R. E., Lightning Spectroscopy, in *Lightning, Vol. 1, Physics of Lightning,* R. H. Golde, editor, Academic Press, N.Y., 1977, pp. 281-306.

chapter twelve

How Is Thunder Generated?

*How Can It Be Used To Measure The Distance
And Length Of The Lightning Channel?*

Throughout recorded history there has been a succession of interesting views on the origin of thunder. Aristotle, in the fourth century B.C., believed that thunder was the noise created when "dry exhalation" forcibly ejected from a cooling cloud struck the surrounding clouds — "dry exhalation", according to Aristotle, being one of the constituents of air (Ref. 12.1). We know now that there is no such substance. Interestingly, Aristotle thought that the origin of thunder preceded that of lightning and that the latter was a "burning wind" produced after the impact of the "dry exhalation" on a surrounding cloud. Lucretius, in the first century B.C., wrote that thunder was produced by clouds banging together (Ref. 12.2). In 1637 Descartes, the mathematician after whom the cartesian coordinate system is named, suggested that thunder was an organ-pipe effect: that thunder was due to a resonance of the air between two clouds occurring when a higher cloud descended upon a lower cloud (Ref. 12.3).

All of these early theories erroneously associated thunder with clouds rather than with the lightning channel. Later speculation regarding the origin of thunder generally associated thunder with the lightning channel. For example, Benjamin Franklin wrote in 1749 (see Chapter 1) that lightning and laboratory sparks were similar in that they both gave a "crack or noise in exploding". The correct meaning for Franklin's "exploding" was not generally agreed upon until the twentieth century. In the mid-nineteenth century the accepted theory seems to have been that lightning produced a vacuum along its path, and thunder resulted from the subsequent motion of air into the vacuum. Two other theories that achieved some respectability involved (1) the chemical explosion of gaseous materials created by the lightning, and (2) the creation of steam explosions along the heated lightning path. One of the first adequate descriptions of the origin of thunder was published by M. Hirn in 1888 (Ref. 12.4).

The thunder we hear is, in fact, created in the following way: The return stroke rapidly deposits a large amount of energy along the leader channel. That channel is heated by the energy input to above $50,000°$ F (see Chapter 11). Heating of a short section (say 10 yards) of channel takes only millionths of a second and hence the channel section has no time to expand while it is being heated. Air heated from room temperature or from a leader temperature of a few thousand degrees to above $50,000°$ F without having time to expand attains a pressure considerably in excess of normal atmospheric pressure (1 atmosphere). The initial pressure of the return stroke channel is definitely in excess of 10 atmospheres and may be 100 atmospheres or more. The high pressure channel rapidly expands into the surrounding air (initially at atmospheric pressure) and compresses it. This disturbance of the air propagates outward in all directions. For the first ten yards or so it propagates as a *shock wave* (a major disturbance in the air which travels faster than the speed of sound) and after that as an ordinary sound wave (small compressions and expansions of the air density). The sound pulse from a short section of lightning channel lasts less than 0.1 seconds and travels at about 1090 ft/second (about 0.21 miles/second) at sea level. The thunder we hear, then, is the pressure variations induced in the air by the expansion of each part of the lightning channel (main channel and branches) due to its initial high pressure.

Since sound travels about a foot in a thousandth of a second and the return stroke heats the entire lightning channel in less than a thousandth of a second (see Chapter 9), for all practical purposes each point on the lightning channel can be considered to emit a pulse of sound (or a shock wave) at the same time. Because light emitted by the return stroke channel travels away from the channel at 186,000 miles/second (or a mile in about 5 millionths of a second), we see light almost simultaneously with channel formation, but the sound of the thunder from that channel frequently takes many seconds to reach us. The initial thunder heard comes from the point on the lightning channel nearest to the observer. The last sound (excluding echoes) comes from the point farthest away. In Fig. 12.1, a 2-mile long intracloud discharge is oriented along the line of sight. We assume (direct evidence is lacking) that sound generation by the intracloud discharge occurs essentially simultaneously at all points on the channel as it does with the ground discharge. From the time it takes for the first sound to arrive, we can establish the distance to the nearest point on the channel. The first sound, traveling about 0.21 miles/second, arrives from the nearest channel point in about 13.5 seconds; multiplying these two figures provides the distance to the

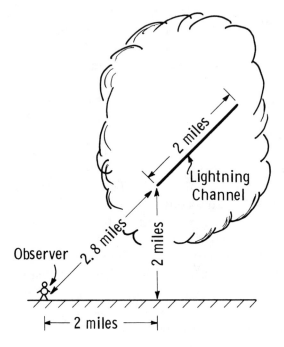

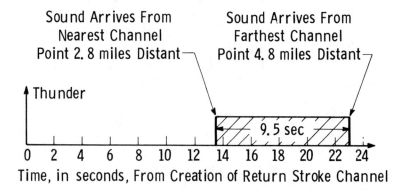

Fig. 12.1. Intracloud lightning flash and its thunder.

nearest point on the channel — 2.8 miles. (In other words, if the time until the thunder arrives is T, the distance to the channel section producing the thunder is roughly T/5, the formula given in Chapter 10.) Sound from the far end of the channel, 4.8 miles away, arrives about 23 seconds after the channel is created, ending the thunder. The duration of the thunder is about 9.5 seconds, from which we can establish the channel length (channel length in miles = 9.5 seconds x 0.21 miles/second). The result represents the actual channel length only if the channel is oriented along the observer's line of sight. Otherwise it is an underestimate of the true length (that is, the true length is longer than the calculated value) as we can see by reference to Fig. 12.1 and 12.2.

In Fig. 12.2, a 2-mile long channel to ground produces sound from its base which arrives at the observer 2 miles from the base in about 9.5 seconds (2 miles = 9.5 seconds x 0.21 miles/second). Sound from the top of the channel, 2.8 miles away, arrives in about 13.5 seconds and thus the total thunder duration is about 4.0 seconds. By measuring only the thunder duration we can conclude that the minimum channel length that could have produced the thunder was 4.0 seconds x 0.21 miles/second = 0.84 miles. The true channel length, 2 miles, is of course longer than the calculated minimum value. Sometimes actual thunder durations are as long as a minute (minimum channel lengths of over 10 miles).

To summarize, from a measurement of the time of arrival of the first sound of thunder we can determine the distance to the closest point of the lightning channel; and from the thunder's duration we can determine a minimum length for the channel. In both cases a distance is derived by multiplying the appropriate time interval by the speed of sound. This is the case even if the channel under consideration is not straight, as the reader may easily prove to himself. It is worth noting that a visible channel to ground may be much farther away than its thunder would indicate if the in-cloud part of the lightning channel passes overhead.

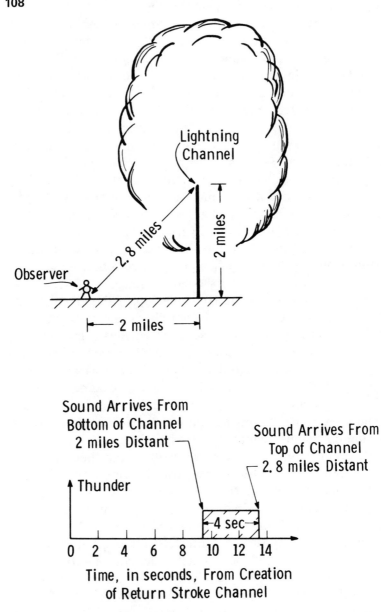

Fig. 12.2. Cloud-to-ground lightning flash and its thunder.

Thunder is generated by each of the return strokes in a multiple-stroke flash. Since the time between the first and last strokes may be a few tenths of a second, the thunder duration may be increased by a few tenths of a second over the time taken for sound to arrive from different ends of the channel. The error made in computing minimum channel lengths due to this effect is not important. It should be noted that the last stroke in a multiple-stroke discharge usually has the greatest length (see Chapter 10), and hence produces the longest lasting thunder. It is therefore the length of the final stroke that is measured.

In addition to thunder produced by return strokes, sound is also produced by stepped leaders and dart leaders. Leader thunder is relatively weak compared with that produced by return strokes.

Why does thunder sound like it does? Why does it rumble, roll, peal, and clap? Two factors must be considered: the intensity or loudness of the sound, and its pitch or frequency. We look first at the loudness. If a section of the tortuous lightning channel is perpendicular to the observer's line of sight, all points on that section will produce sound that arrives almost simultaneously at the observer, and the result is a high-intensity sound — a peal or a clap. If a channel section is along the observer's line of sight (as with the whole channel in Fig. 12.1), the sound arrival times from various points on the channel will be spread out, resulting in a low-intensity sound — a rumble or a roll. Channel tortuosity enhances the rumbling effect. Typical thunder consists of rumble and roll on which three or four peals or claps are superimposed.

Now to pitch. As we have seen, thunder generation is similar to an explosion. As everyone knows, small explosions (e.g., small firecrackers) make a high pitched sound while large explosions (e.g., large aerial fireworks or sticks of dynamite) make a low pitched sound. The larger the energy input, the lower the pitch. Thunder generally has a pitch, or frequency, of about 50 cycles per second. From measurements of thunder pitch and use of appropriate theory, scientists have determined the energy input per unit length of lightning stroke channel (100,000 to 1,000,000 watt-seconds per yard of channel length). Thunder pitch is influenced by two additional factors: First, the lower the air density for a given energy input to the lightning channel, the lower the thunder pitch; this means that thunder from a channel at high altitude should have a lower pitch than thunder from a channel near ground. Second, air filters out the higher pitches preferentially, so that the further away thunder is heard from its channel, the lower should be its pitch.

Thunder is seldom heard more than 15 miles away from a lightning channel. The reason is that, due to the effects of temperature and wind variation with height, sound waves generally curve upward in the air under and surrounding thunderstorms. Thus, the thunder from a lightning channel passes over the head of an observer if he is more than about 15 miles from the channel. If he climbed a tall enough ladder, he would hear the thunder.

Within a few hundred yards of a lightning strike, the observer hears one loud (terrifying) bang possibly preceded by hissing or clicking or something similar to cloth tearing. The bang is the sound wave from the channel base. No long-duration thunder is usually heard because, due to the upward curvature of the sound waves, the thunder from sections of the channel above the channel base pass over the observer's head. The various sounds reported to occur before the loud bang may be due to the downward-moving stepped leader, to upward-moving discharges attempting to meet the stepped leader, and/or to a general corona-discharge blanket at the ground.

REFERENCES

12.1. Aristotle (384-322 B.C.), *Meteorologica*, H. P. D. Lee, trans., Loeb Classical Library, Harvard University Press, Cambridge, Mass., 1951, pp. 223-225.

12.2. Lucretius, T. (98-55 B.C.), *On the Nature of Things*, book VI, H. A. J. Munro, trans., Great Books of the Western World, William Benton, Chicago, 1952, p. 81.

12.3. Haldane, E. S., *Life of René Descartes,* John Murray (Publishers), Ltd., London, 1905, p. 181.

12.4. Hirn, M., The Sound of Thunder, *Sci. Am., 59,* 201 (1888).

A general reference for this chapter and one which contains a more detailed bibliography of the early theories of thunder is: *Lightning,* M. A. Uman, Dover Publications, Inc., New York, 1984, Chapter 6.

ADDITIONAL READING

Few, A. A., Acoustic Radiations from Lightning, in *CRC Handbook of Atmospherics, Vol. 2,* H. Volland, editor, CRC Press, Boca Raton, Florida, 1982, pp. 257-290.

chapter
thirteen

Does Lightning Occur Without Thunder?
Thunder Without Lightning?

What Is Heat Lightning? Sheet Lightning?

Ask a lot of people the question "Did you have lightning in your vicinity yesterday?" and, assuming there was lightning, at least one person will invariably answer "No lightning, but a lot of thunder!" Possibly, this is just another way of saying that thunder was heard but lightning wasn't seen. More likely, the answer indicates that the person does not understand the relation between lightning and thunder.

We have seen in the previous chapter that the origin of thunder is the explosive expansion of the lightning channel following a rapid energy input to the channel. Since it is the lightning channel that creates the thunder, there can be no thunder without lightning. But can there be lightning without thunder? Strictly speaking, no. Every electrical discharge, when initiated, produces some noise. However, if we ask whether there is lightning whose thunder may be inaudible a relatively short distance from the channel, the answer is apparently yes. There have been lightning flashes reported to strike the Washington Monument without producing audible thunder, and flashes to the Empire State Building which were reported to have produced a sound described as being the same as if a paper bag full of water had been dropped a short distance to the floor (Ref. 13.1).

As noted in Chapter 6, lightning flashes to tall buildings are generally initiated by stepped leaders which move upward from the building tops. The current in a structure-initiated flash rises slowly to a peak current of a few hundred amperes and this low level of current typically flows for a few tenths of a second. Often this continuous current is punctuated by current peaks caused by return strokes whose leaders are initiated in the cloud and travel from cloud to building top down the active channel. Sometimes no return strokes occur and the flash consists of only the low-level current. It is the upward-initiated discharge with no return strokes that might be expected to generate very little sound. Its current is small and increases slowly compared to the current of a typical return stroke; the energy input is neither large nor rapid and thus cannot lead to the explosive channel expansion and typical thunder characteristic of the usual return stroke channel.

What about "heat lightning"? Why does it have no thunder? Before we answer this question, we had better define what we mean by "heat lightning". In common terminology, heat lightning is the name given to the illumination of distant clouds close to the horizon which occurs without thunder and in the absence of a visible lightning channel. Heat lightning apparently is so named because it occurs on hot nights. The fact that it occurs on hot, generally humid nights is due simply to the fact that hot humid air near the ground is an ingredient in the thunderstorm recipe (see Chapter 7). The nocturnal thunderstorms accused of generating heat lightning actually produce the usual intracloud and cloud-to-ground lightning flashes. The light from these flashes illuminates the clouds from within and by reflection so that the clouds are easily seen even when they are sufficiently far away (say 30 miles) that individual lightning channels cannot be discerned. Since thunder cannot generally be heard more than about 15 miles from a lightning channel (see Chapter 12), the distant heat lightning produces no audible thunder.

"Sheet lightning" is a close relative of heat lightning. Sheet lightning is the name given to the intracloud discharges which light up a large cloud area simultaneously, giving the impression of a sheet of light. Sheet lightning may or may not produce audible thunder, depending on its distance from the observer.

REFERENCES

13.1.   McEachron, K. B., Lightning to the Empire State Building, *J. Franklin Institute, 227,* 149-217 (1939).

chapter
fourteen

Do Gushes Of Rain Follow Thunder?

Observers of thunderstorms, from antiquity to the present, have noted that a heavy gush of rain often reaches ground a minute or two after a lightning flash and its accompanying thunder (Refs. 14.1, 14.2). Since precipitation is believed necessary to provide a thunderstorm with its electrical charge (see Chapter 8), it could be argued that a rain gush occurring in the cloud provides the charge separation that causes the lightning. On the other hand, it could just as well be argued that the lightning or its thunder modifies the character of the precipitation in the cloud in such a way as to cause a rain gush. Finally, it is also possible that both lightning and the rain gush are caused by some other effect or that it is only fortuitous that they appear to be related. Not until the 1960s did radar observations of the precipitation in clouds before and after lightning show that lightning and thunder precede the formation of the rain gush (Refs. 14.1, 14.2).

As long ago as 58 B.C., Lucretius wrote that gushes of rain were caused by thunder. Modern proponents of this currently unpopular view suggest that the compression and rarefaction waves associated with the thunder accelerate some drops relative to others causing collisions between the drops which would otherwise not have occurred. The collisions allow small drops to coalesce into bigger drops which are heavy enough to fall out of the cloud as rain. It is interesting to note that from radar measurements of clouds there is evidence to suggest that thunder can cause supercooled water droplets (water droplets below $32°F$ which are not frozen) in the clouds to turn to ice (Ref. 14.3). Further, supercooled water droplets can be turned to ice crystals in the laboratory by the sound from a toy pop gun or by a bursting balloon (Ref. 14.4). The freezing mechanism involved in both the laboratory and the cloud may be the lowering of the droplet temperature by the expansion of the air associated with the sound wave rarefaction.

Most investigators now believe that the lightning/rain gush relation is electrical rather than acoustic. As seen in Chapters 8 and 9, the top of the lightning channel is found in the N-region of the cloud. The N-charge resides on supercooled water and ice particles. Stepped and dart leaders are negatively charged, drawing their charge from the N-region. The return stroke drains the negative charge on the leader to the ground. After the return stroke has reached the top of the lightning channel, the whole channel, a good conductor tied to ground, becomes positively charged in response to the negative cloud charge above and around it. The leader-return stroke process results in a negatively-charged leader channel being replaced by a positively-charged channel in a time that is short compared to that in which any of the surrounding negative cloud charge can move appreciably (Fig. 9.3a). If electrical effects cause the rain gush, the charged cloud particles must coalesce in response to the rapid change in the sign of the charge on the lightning channel.

The mechanism for coalescence most closely examined is that proposed by Bernard Vonnegut and Charles Moore (Refs. 14.1, 14.2, 14.5). They postulate that the cloud droplets in the immediate vicinity of the newly-completed return stroke channel acquire a net positive charge from the positively-charged channel. Since similar charges repel one another, the positively charged droplets fly away from the channel, initially at perhaps tens of yards per second, colliding and coalescing with the surrounding negatively-charged droplets. The initially-stationary negative droplets are attracted toward the moving positive droplets by virtue of their opposite charge, further increasing the collision rate. The positive droplets may travel about 10 yards, increasing in size all the time. The end result is a group of large water drops surrounding the channel.

If the process just described is to produce a rain gush, it must be operative over a reasonably large volume of the cloud. Vonnegut and Moore suggest that this is the case because the lightning channel in the cloud has a tree-like structure. A drawing of this structure is shown in Fig. 9.3. Some idea of the extent of the structure may be inferred from electrical discharges in insulating materials (the cloud is essentially an insulator). In Fig. 14.1 a photograph of a plastic block in which a tree-like discharge has taken place is shown. The block was charged negatively by shooting high energy electrons into it. The tree-like discharge occurred when a grounded electrode was placed at the edge of the block. Unfortunately, there is at present no direct evidence regarding the extent of the lightning channel structure in the cloud, and we must await future lightning research to determine its actual structure.

Some of the gushes of precipitation that fall after lightning are in the form of hail (Ref. 14.2). When a rain gush forms by a coalescence process at temperatures below $32°F$, the drops may freeze as they become large and then continue their growth as hail particles.

While most rain gushes are preceded by lightning, it does not follow that lightning invariably produces rain gushes. Whether there is a rain gush or not depends, among other things, on the liquid water content of the cloud. If this is too low, heavy rain cannot form.

Although it is possible for lightning to occur in the absence of appreciable precipitation, the two usually go hand in hand. Recently the amount of rainfall reaching the ground near Tucson, Arizona has been correlated with the number of cloud-to-ground discharges (Ref. 14.6). These results and those of earlier studies show that the greater the number of lightning discharges, the greater the rainfall. (The clouds which produce both the most rain and the most lightning are those of the greatest vertical extent.) From measurements of

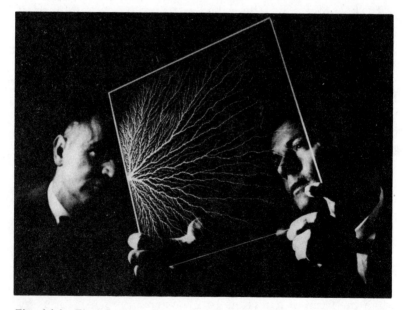

Fig. 14.1. Electrical breakdown in a plastic block. The tree-like structure which extends into the block may be similar to the lightning discharge structure extending from the top of the channel upward into the cloud. (Courtesy, Arthur D. Little, Inc.)

the amount of rainfall and the number of lightning flashes occurring near Tucson between 1 and 6 p.m. on 52 days, it was determined that an average of about a thousandth of an inch of rain arrives at the ground for each cloud-to-ground lightning stroke observed. On 23 days of heavy rain (exceeding 0.10 in.) there was an average of 175 cloud-to-ground lightning flashes per day. On 29 days of light rain (less than 0.01 in.) there was an average of only three flashes per day.

REFERENCES

14.1. Moore, C. B., B. Vonnegut, J. A. Machado, and H. J. Survilas, Radar Observations of Rain Gushes Following Overhead Lightning Strokes, *J. Geophys. Res., 67*, 207-220 (1962).

14.2. Moore, C. B., B. Vonnegut, E. A. Vrablik, and D. A. McCaig, Gushes of Rain and Hail after Lightning, *J. Atmosph. Sci., 21*, 646-665 (1964).

14.3. Vonnegut, B., and C. B. Moore, Nucleation of Ice Formation in Supercooled Clouds as the Result of Lightning, *J. Appl. Meteorology, 4*, 640-642 (1965).

14.4 Vonnegut, B., Production of Ice Crystals by Adiabatic Expansion of Gas, *J. Appl. Phys., 19*, 959 (1948).

14.5. Vonnegut, B., and C. B. Moore, A Possible Effect of Lightning Discharge on Precipitation Formation Process, *Am. Geophys. Union Monograph Number 5*, 287-304 (1960).

14.6. Battan, L. J., Some Factors Governing Precipitation and Lightning from Convective Clouds, *J. Atmosph. Sci., 22*, 79-84 (1965).

ADDITIONAL READING

Piepgrass, M. V., and E. P. Krider, Lightning and Surface Rainfall During Florida Thunderstorms, *J. Geophys. Res., 87*, 11,193-11,201 (1982).

Szymanski, E. W., S. J. Szymanski, C. R. Holmes, and C. B. Moore, An Observation of Precipitation Echo Intensification Associated with Lightning, *J. Geophys. Res., 85*, 1951-1953 (1980).

Zrnic, D. S., W. D. Rust, and W. L. Taylor, Doppler Radar Echoes of Lightning and Precipitation at Vertical Incidence, *J. Geophys. Res., 87*, 7179-7191 (1982).

chapter fifteen

What is Ball Lightning?

Are UFOs and Ball Lightning Related?

Ball lightning (in French, *boules de feu* or *foudre sphérique;* in German, *Kugelblitz*) is the name given to the mobile luminous spheres which have been observed during thunderstorms. A typical ball lightning is about the size of an orange or a grapefruit and has a lifetime of a few seconds. A 19th century woodcut depicting the occurrence of a somewhat larger than average ball lightning is shown in Fig. 15.1. A number of compilations of eye-witness reports of ball lightning have been published (Refs. 15.1 to 15.6). Visual sightings are often accompanied by permanent material damage, sounds, and odors. Nevertheless several noted scientists have attributed ball lightning to persistence of vision (after seeing a bright lightning channel) or to other optical effects (Ref. 15.7).

In a letter to the editor of the London *Daily Mail* (Ref. 15.8) W. Morris described an unusual incident in which a ball lightning caused a tub of water to boil:

● During a thunderstorm I saw a large, red hot ball come down from the sky. It struck our house, cut the telephone wire, burnt the window frame, and then buried itself in a tub of water which was underneath.

The water boiled for some minutes afterwards, but when it was cool enough for me to search I could find nothing in it.

As a researcher in the field of lightning, the writer has personally received over one hundred unsolicited eye-witness accounts of ball lightning. Here are five representative observations:

Fig. 15.1. Ball lightning in a 19th century woodcut. The original title, translated from French, reads "Ball lightning crossing a kitchen and a barn." Perhaps the ball lightning came down the chimney used to exhaust the cooking fires. How the young lady's blouse came to be in such a state of disarray is not known. (Courtesy, Burndy Library)

● I was visiting my uncle's farm in Eastern Norway where the country is very flat and the summers warm with cloud gatherings and thunderstorms in the middle of the day. One day when my cousin and I went shopping, a thunderstorm worse than usual broke out. The rain poured down, and we stayed in the open door of the shop, looking out. Suddenly we saw a glowing ball, the size of a small football [soccerball], passing the door at great speed a little above ground level. We had never heard about 'ball lightning' before, and wondered what it might be. When the rain stopped, we started on our way home and walked along a path leading through the forest. Then it started thundering again, and another ball, just like the first one, came rolling towards us. Now we got curious, but not very much — to 10-year-old children there are so many strange things to wonder about — so what is a rolling glowing ball more or less? Then my uncle came to meet us, because he thought we might be frightened by the heavy thundering, and the first thing he said was, 'Did you see the ball lightning?' Not till then did we realize that we had seen something strange and dangerous.

● While outdoors during a particularly violent thunderstorm such as are common there [Iowa], I heard a heavy rushing noise like an extra strong wind. This caused me to look about me and I saw a ball of fire, yellowish white about the size of a wash tub bouncing down the dirt road. The ball traveled a little faster than one could run. I did not see from where it came. I watched the ball travel about a city block, when it struck a small shed, maybe 10 feet by 12 feet in which a horse was stabled. The shed seemed to explode and the horse was killed.

● Near Murray's dairy, a property on the Queanbeyan road on the then outskirts of Canberra [Australia], and not far from what is now the industrial area of Fyshwick, I was riding along the right hand side of the road, just off the paving to prevent a possible slip and fall on the wet surface, and about 20 to 25 yards ahead, on the left hand side and also off the bitumen, a farm employee was leading a Shorthorn bull. There was very little, if any rain at that time. I can clearly recall that there was one of those periods of "quietness" that sometimes precedes a downpour. Just as I drew level with the bull there was one very loud bang or explosion and immediately down the white traffic line in the centre of the road appeared the fireball. It seemed to be about 6" or 8" off the ground, was about the size of a basketball like very golden butter in color, and had the appearance of being "spun" or "fuzzy", like silk threads or wool, as distinct from a "molten" liquid look. It did not sparkle — just a ball of fuzz. It came straight down say three of the white marker lines. These have about equal distance of unmarked road between them, so I should say it travelled about 18 to 20 feet. Then it simply disappeared. It did not break apart. There was no further noise like an explosion. It was there one moment and not there the next.

The whole thing would have been over in probably two or three seconds, before the horse had time to be startled.

The young men leading the bull cried out — in Pure Australian — "What the bloody hell was THAT?".

As there was another downpour shortly, we did not stop to discuss it.

● I was standing in the kitchen of my home in Omaha, Nebraska while a terrible thunderstorm was in progress. A sharp cracking noise caused me to look toward a window screen to my left. Then I saw a round, iridescent (mostly blue) object, baseball size, coming toward me. It curved over my head and went through the isinglass [mica] door of the kitchen range, striking the back of the oven and spattering into brilliant streamers. There was no sound and no effect on me except a tingle as it passed over my hair. Later examination showed a tiny hole with scorched edges in the screen and isinglass, and scorch-like marks on the back of the oven.

● The whole family was sitting at the supper table when a ball about 4 inches in diameter came and fluttered about the center of the table. It made a buzzing noise and was about 8 to 10 inches above the dishes. Its color was a mixture of blue and orange and may have had some red. It fluttered about 5 to 6 seconds. I could easily have grabbed it had I dared to do so. Then it exploded with a loud bang like a large fire cracker and gave off a smell somewhat like city cooking gas which lasted several minutes. After it exploded the table was left as before, no dishes broken or moved about.

A number of photographs purported to be of ball lightning have been published (Refs. 15.9 to 15.14). Most of them are time exposures in which the ball lightning appears as a meandering ribbon of light. Some of the photographs may well be of automobile or other moving lights, or of stationary lights photographed by a moving camera. It appears that no motion pictures of ball lightning have been taken.

A phenomenon very similar to, if not identical with, ball lightning has been reported to occur in submarines due to the discharge of a high current (about 150,000 amp D.C. from a 260 volt source) across a circuit breaker (Ref. 15.15). In addition, the writer has received a number of reports of ball-lightning-like phenomena being initiated in electrical equipment, both with and without lightning present. One such thunderstorm-associated incident was reported by a retired professor and engineering department chairman at a major state university:

● In 1914 I was working for the Interurban Co. and happened to be in the main power house during a very severe thunder storm. We had recently installed a 600 volt dc rotary converter which supplied power to both the city and interurban lines. I was working on the main switch board and watching a bit carefully as there were many breakers opening. I looked up just in time to see a ball of fire about 15 to 18 inches in diameter come off the converter commutator and float at a good speed down the length of the switch board, maybe three feet above my head. Some hundred feet from the converter it struck the ceiling of an office room where a cat was asleep. The ball splashed in all directions like water and the cat sprang high in the air. The cat squalled loudly but seemed to be uninjured. In both cases there was the strong odor of ozone, but due to the constant lightning I would not attribute it to the 'ball lightning.'

Ball lightning has been seen by 5 to 10% of the population. Although it is generally thought that ball lightning is a rare phenomenon, recent research has shown that the numbers of ball lightning observers are not much different from the numbers who have observed lightning impact points (Ref. 15.6). The implication is that ball lightning is usually created at or near the lightning channel and that an appreciable fraction of all cloud-to-ground lightning flashes may give birth to ball lightning. Since the balls generally last for only a few seconds, they cannot get too far from the mother channel. Thus ball lightning may well be common, but rarely seen.

Ball lightning and St. Elmo's fire are sometimes confused. St. Elmo's fire is a corona discharge induced at a pointed conducting object by thunderstorm charges. Like ball lightning, St. Elmo's fire may assume a spherical shape. Unlike ball lightning, it must remain attached to a conductor, although it may exhibit some motion along the conductor. Further, St. Elmo's fire may last much longer than the typical ball lightning.

From the many published ball lightning observations, it is possible to compile a list of ball lightning characteristics:

Occurrence: Most observations of ball lightning are made during thunderstorm activity. Most, but not all, of thunderstorm-related ball lightning appears almost simultaneously with a cloud-to-ground lightning discharge. These balls appear within a few yards of the ground. Sometimes ball lightning is reported to appear near ground in the absence of a lightning discharge. Lightning balls have also been observed to hang in mid-air above the ground and have been observed falling from a cloud toward the ground.

Appearance: Ball lightning is generally spherical, although other shapes have been reported. Balls are usually 4 to 8 in. in diameter, with reported diameters ranging from 1/2 in. to many feet. Various colors have been observed, the most common being red, orange, and yellow. Luminosity is generally not exceptional but the balls can be seen clearly in daylight. They are usually reported to maintain a relatively constant brightness and size, although changes in brightness and size are not uncommon.

Lifespan: Ball lightning generally lasts less than 5 seconds. In a small fraction of the reports a lifespan of over a minute is indicated.

Motion: Lightning balls usually move horizontally at a velocity of a few yards per second. They may also remain motionless in mid-air or may descend from a cloud toward the ground. They do not often rise, as would be the case if they were spheres of hot air at atmospheric pressure subject only to gravity. Many reports describe balls which appear to spin or rotate as they move. Sometimes they are reported to bounce, typically off the ground.

Heat, sound, and odor: Rarely is the sensation of heat reported although there are accounts of lightning balls which burned barns and melted wires. One report (Ref. 15.5) describes a ball lightning which hit a pond of water with a sound "as if putting a red hot piece of iron into the water." Sometimes a hissing sound is said to be emitted. Many observers report an accompanying odor usually described as sharp and repugnant, resembling ozone, burning sulphur, or nitric oxide.

Attraction to objects and enclosures: Reports often describe balls attached to, and moving along, metallic objects such as wire fences or telephone lines. Some or all these observations may refer to a type of St. Elmo's fire. Lightning balls often enter houses through screens or chimneys, and sometimes through glass window panes. They are also reported to originate within buildings, on occasion from telephones. Balls can exist in an all-metal enclosure such as the interior of an airplane. For instance:

● I was at the controls of a KC-97 USAF tanker aircraft, heavily loaded with JP-4 fuel for offload to B-47 bombers. En route to the refueling rendezvous (Elko, Nevada vicinity) we were in the clouds at 18,000'. There was light precipitation, temp. was above freezing and there was no turbulence.

I recall that St. Elmo's fire was dancing around the edges of the aircraft front windows. (This is a not too uncommon occurrence but may have some significance to you.) The crew was experienced in all phases of all-weather operation and not concerned or apprehensive about any portion of the mission to be accomplished.

As I was concentrating on the panel (no outside visual references were visible) a ball of yellow-white color approximately 18″ in diameter emerged through the windshield center panels and passed at a rate about that of a fast run between my left seat and the co-pilot's right seat, down the cabin passageway past the Navigator and Engineer. I had been struck by lightning 2 times through the years in previous flights and recall waiting for the explosion of the ball of light! I was unable to turn around and watch the progress of the ball as it proceeded to the rear of the Aircraft, as I was expecting the explosion with a full load of JP-4 fuel aboard and concentrated on flying the aircraft. After approximately 3 seconds of amazingly quiet reaction by the 4 crew members in the flight compartment, the Boom operator sitting in the rear of the aircraft called on the interphone in an excited voice describing a ball of fire that came rolling through the aft cargo compartment abeam the wings, then danced out over the right wing and rolled off into the night and clouds! No noise accompanied the arrival or departure of the phenomenon. (Ref. 15.16)

● This communication records the observation of ball lightning in unusual circumstances. I was seated near the front of the passenger cabin of an all-metal airliner (Eastern Airlines Flight EA 539) on a late night flight from New York to Washington. The aircraft encountered an electrical storm during which it was enveloped in a sudden bright and loud electrical discharge (0005 h EST, March 19, 1963). Some seconds after this a glowing sphere a little more than 20 cm [8 inches] in diameter emerged from the pilot's cabin and passed down the aisle of the aircraft approximately 50 cm [20 inches] from me, maintaining the same height and course for the whole distance over which it could be observed.

. . . the relative velocity of the ball to that of the containing aircraft was 1.5 ± 0.5 meters [or yards] per second . . . the object did not seem to radiate heat. . . . the optical output could be assessed as 5 to 10 watts and its color was blue-white. . . . the course was straight down the whole central aisle of the aircraft. (Ref. 15.17)

Demise: Ball lightning decays in one of two modes, either explosively or silently. Explosive decay occurs rapidly and is accompanied by a loud noise. Silent decay can take place either rapidly or slowly. Most lightning balls apparently exhibit a rapid decay. After the ball has decayed, it is sometimes reported that a mist or residue remains. Occasionally a ball lightning has been observed to break up into two or more smaller balls.

Types: There may be more than one type of ball lightning. For example, the ones that attach to conductors may differ from the free-floating balls; the balls that appear near ground level may differ from those that hang high in the air or those that fall out of a cloud.

Unfortunately, there is at present no adequate theory of ball lightning. For example, no theory can account simultaneously for the degree of mobility, the constancy of light output, and for the fact that the ball does not rise. Despite numerous · theoretical models proposed, the causal mechanisms remain unknown. All theories fall into one of two general classes: those in which the energy source for the ball is postulated to be outside the ball (externally powered ball lightning) and those in which the sustaining energy is postulated to be stored within the ball itself (internally powered ball lightning).

In the internally powered models there are essentially six subclasses: (1) Ball lightning is gas or air behaving in an "unusual" way. It has been suggested that the ball lightning is slowly burning gas, is the light produced by the slow recombination of unspecified ions existing in the ball, is due to chemical reactions involving dust, soot, etc., and so on. (2) Ball lightning is a sphere of heated air or heated air containing various impurities at atmospheric pressure – Ref. 15.18. (3) Ball lightning is a very high-density plasma (ionized gas) which exhibits properties characteristic of solid materials – Ref. 15.19. (4) Ball lightning is due to one of several suggested configurations of closed-loop current flow contained by its own magnetic field. It has been shown that plasma containment of this type is not possible in air — Ref. 15.20. (5) Ball lightning is due to some sort of air vortex (like a smoke ring) providing containment for luminous gases. (6) Ball lightning is a high frequency electromagnetic field contained within a thin spherical sheet of ionized air — Ref. 15.21.

In the theories which provide the ball with an external power source, three types of power sources have been suggested: (1) A high-frequency (hundreds of megacycles per second) electromagnetic field, (2) a steady current flow from cloud to ground, and (3) focused cosmic-ray particles. (1) M. Cerrillo (Ref. 15.22) and P. Kapitza (Ref. 15.23) proposed that focused radio frequency energy from the thundercloud could create and maintain a ball lightning. Radio waves with the energy necessary to effect this mechanism have never been observed in thunderstorms. (2) D. Finkelstein and J. Rubinstein (Ref. 15.20), the writer and C. Helstrom (Ref. 15.24), and J. Powell and Finkelstein (Ref. 15.25) have suggested that a steady current flowing from cloud-to-ground would contract in cross section at the ball (originally provided by part of the lightning channel) and that the increased energy input due to the constriction of current could maintain the ball. This type of theory cannot account for the existence of ball lightning inside structures, particularly inside metal structures. (3) V. Arabadzhi (Ref. 15.26) has suggested that radioactive cosmic-ray particles could be focused by the charges of the thundercloud so that the cosmic rays would create an air discharge at one point in space. This would appear to be very unlikely.

Are UFOs (Unidentified Flying Objects) and ball lightning related? There are all kinds of UFOs. Some are hoaxes; some are misinterpretations of well-understood physical phenomena like meteors, not known to the observer or seen under unusual circumstances; some, like re-entry vehicles, are unpublicized objects of our advanced technology; some are imperfectly understood physical phenomena; some may be due to poorly understood psychological effects; and, finally, some may be spaceships (true flying saucers). Although there are many observations of flying-saucer-like objects reported by reliable people (Refs. 15.27, 15.28), incontrovertible proof that interplanetary spaceships exist — a captured or crashed saucer — has not been forthcoming. The UFO literature referenced (Refs. 15.27, 15.28) and the references contained in that literature make fascinating reading.

But let us return to our original question. A small percentage of the UFO reports are so similar to a certain class of ball lightning reports that they both must refer to the same "imperfectly understood physical phenomena." These particular lightning balls are much larger, much brighter, and much longer-lived than the typical balls. They are reported to be 10 to 20 ft in diameter, give the impression of being as bright as lightning, and may last a minute or more. When such objects appear immediately after lightning, it is clear that they should be called ball lightning. Sometimes, however, they appear near or in clouds or in snow without the apparent presence of lightning, and occasionally such objects are reported in seemingly clear air.

REFERENCES

15.1. Brand, W., *Der Kugelblitz*, Grand, Hamburg, Germany, 1923.

15.2. Rodewald, M., Kugelblitzbeobachtungen, *Z. Meteorol.*, *8* 27-29 (1954).

15.3. Dewan, E. M., Eyewitness Accounts of Kugelblitz, Microwave Physics Laboratory, Air Force Cambridge Res. Labs., *CRD-125*, March 1964.

15.4. Silberg, P. A., A Review of Ball Lightning, *Problems of Atmospheric and Space Electricity*, S. C. Coroniti (ed.), pp. 436-454, American Elsevier Publishing Company, New York, 1965.

15.5. NcNally, J. Rand, Jr., Preliminary Report on Ball Lightning, Oak Ridge Natl. Lab., *ORNL-3938, UC-34-Phys.*, May 1966. Rayle, W. D., Ball Lightning Characteristics, *NASA Technical Note D-3188*, January 1966.

15.6. Rayle, W. D., Ball Lightning Characteristics, *NASA Technical Note D-3188*, January, 1966.

15.7. Humphreys, W. J., Ball Lightning, *Proc. Am. Phil. Soc.*, *76*, 613-626 (1936).

15.8. Morris, W., A Thunderstorm Mystery, letters to the editor of *Daily Mail* of London, Nov. 5, 1936.

15.9. Jensen, J. C., Ball Lightning, *Physics* (now *J. Appl. Phys.*), *4*, 372-374 (1933).

15.10. Kuhn, E., Ein Kugelblitz auf einer Moment-Aufnahme?, *Naturwissenschaften, 38*, 518-519 (1951).

15.11. Wolf, F., Interessante Aufnahme eines Kugelblitzes, *Naturwissenschaften, 43*, 415-417 (1956).

15.12. Davidov, B., Rare Photograph of Ball Lightning, *Priroda, 47*, 96-97 (1958). In Russian.

15.13. Jennings, R. C., Path of a Thunderbolt, *New Scientist, 13* (no. 270), 156, January 18, 1962.

15.14. Müller-Hillebrand, D., Zur Frage des Kugelblitzes, *Elektrie, 17*, 211-214 (1963).

15.15. Silberg, P. A., Ball Lightning and Plasmoids, *J. Geophys. Res.*, *67*, 4941-4942 (1962).

15.16. Uman, M. A., Some Comments on Ball Lightning, *J. Atmospheric Terrest. Phys., 30*, 1245-1246 (1968).

15.17. Jennison, R. C., Ball Lightning, *Nature, 224,* 895 (1969).

15.18. Lowke, J. J., M. A. Uman, and R. W. Liebermann, Toward a Theory of Ball Lightning, *J. Geophys. Res.*, *74*, 6887-6898 (1969).

15.19. Neugebauer, T., Zu dem Problem des Kugelblitzes, *Z. Physik, 106*, 474-484 (1937).

15.20. Finkelstein, D., and J. Rubinstein, Ball Lightning, *Phys. Rev., 135*, A390-A396 (1964).

15.21 Dawson, G. A., and R. C. Jones, Ball Lightning as a Radiation Bubble, *Pure and Applied Geophysics, 75*, 247-262 (1969).

15.22. Cerrillo, M., Sombre las posibles interpretaciones electromagneticas del fenomena de las centallas, *Comision Impulsora Coordinadora Invest. Cient., Mexico, Ann., 1*, 151-178 (1943).

15.23. Kapitza, P., The Nature of Ball Lightning, *Dokl. Akad. Nauk SSSR, 101*, 245-248 (1955). In Russian.

15.24. Uman, M. A., and C. W. Helstrom, A Theory of Ball Lightning, *J. Geophys. Res., 71*, 1975-1984 (1966).

15.25. Powell, J. R., and D. Finkelstein, Ball Lightning, *American Scientist, 58*, 262-279 (1970).

15.26. Arabadzhi, V. I., The Theory of Atmospheric Electricity Phenomena, *Uch. Zap. Minsk. Gos. Univ., im A. M. Gor'kogo, Ser. Fiz.-Mat.*, no. 5, 1957. (Translation available as RJ-1314 from Associated Technical Services, Inc., Glen Ridge, New Jersey.)

15.27. *Scientific Study of Unidentified Flying Objects*, E. U. Condon, Scientific Director, D. S. Gillmor, Editor, Bantam Books, New York, 1969. All of the eyewitness reports contained in this book should be read before reading the Conclusions and Recommendations of Section I. Many scientists have arrived at different conclusions from those stated in the book on the basis of the evidence presented in the book.

15.28. *Symposium on Unidentified Flying Objects*, Hearings Before the Committee on Science and Astronautics, U. S. House of Representatives, Ninetieth Congress, Second Session, July 29, 1968, Number 7, U. S. Government Printing Office, Washington, D. C., 1968.

ADDITIONAL READING

Barry, J. D., *Ball Lightning and Bead Lightning,* Plenum Press, New York, 1980.

chapter
sixteen

What is Ribbon Lightning? Bead Lightning?

Ribbon lightning is the name given to the optical illusion occurring when a cloud-to-ground lightning flash is moved sideways an appreciable distance by the wind during the time between the component strokes of the flash. Each stroke in the flash is then seen separated horizontally in space. To the eye, each identically-shaped stroke (ribbon) appears to occur simultaneously. A photograph of ribbon lightning is shown in Fig. 16.1.

The first solid evidence that lightning flashes are generally composed of a number of separate strokes was provided in the 1880s by photographs of ribbon lightning. By the early twentieth century several investigators had obtained ribbon-like photographs of normal (not wind-blown) lightning by moving their cameras as the flash was occurring. Each component stroke of the flash appeared at a different

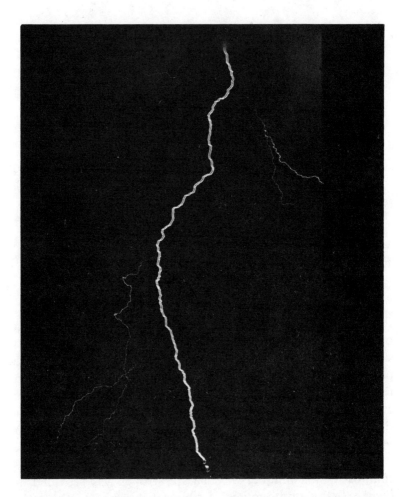

Fig. 16.1. Ribbon lightning near Tucson, Arizona photographed with a tripod-mounted camera. (Courtesy, George Marcek, Catalina High School, Tucson)

position on the film because the lens and film position had changed relative to the fixed lightning position during the time between strokes. The technique of moving (by hand or mechanically) a camera during a flash to time-resolve the features of the flash has yielded much valuable information about lightning. For example, in this way stepped and dart leaders were discovered and their velocities measured. Modern "streak" cameras which time-resolve various high-speed luminous events (including lightning) employ the same basic idea, relative motion between lens and film, as the hand-moved camera. The streak camera lens ·remains stationary while the motor-driven film moves at high speed horizontally behind the lens. A modern streak camera photograph of a 12-stroke lightning flash is shown in Fig. 16.2. A streak camera photograph of a stepped leader and the resultant return stroke is shown in Fig. 16.3. As is evident, the stepped leader channel is photographically dark between steps and elongates itself by a new step each time it lights up. A detailed description of the stepped leader is given in Chapter 9 and the stepped leader's movement towards ground is illustrated in Figs. 9.1 and 9.2.

Bead or chain lightning is a visually well-documented phenomenon in which the lightning channel to ground breaks up, or appears to break up, into luminous fragments generally reported to be some tens of yards long. The beads are reported to persist longer than the usual cloud-to-ground discharge channel.

No reliable still photograph of bead lightning has been published in the literature. Further, it is not clear that a still camera would record a beaded effect. If the channel is first continuous and then breaks up into beads, the image on the film will be a superposition of the continuous channel and the beads. The beads will only be apparent if the light output of the beads to which the film is sensitive is at least a small fraction of the light output of the channel prior to the formation of beads and if the film is not over-exposed.

Fig. 16.2. Streak-camera photograph of a 12-stroke lightning flash taken near Socorro, New Mexico. The first stroke is on the left and is the only branched stroke (see Chapter 9). Increasing time goes from left to right. Continuing current (see Chapter 5), as evidenced by continuing luminosity, flows after the eleventh stroke. (Courtesy, Marx Brook, New Mexico, Institute of Mines and Technology)

To document bead lightning adequately, a movie is probably necessary. Figure 16.4 is a frame from one such movie taken by U. S. Navy personnel while photographing a plume of water rising from a depth charge explosion (Ref. 16.1). The lightning flash of Fig. 16.4 which struck the depth charge plume consisted of three strokes, the first lasting more than 0.5 second, the second about 0.1 second, and the last more than 0.1 second. Continuing current (see Chapter 5) was apparently involved in each stroke. The latter frames of each of the three strokes show a channel which has broken up into light and dark sections a few yards in length.

There are various theories of bead lightning and some of these are now considered:

(1) The visual appearance of bead lightning is caused by

Fig. 16.3. Streak-camera photograph of a stepped leader and the resultant return stroke taken near Lugano, Switzerland. Increasing time goes from left to right; the time duration of the photograph is about 0.002 sec. The vertical height shown is about 400 yards. On the left side of the photograph the intensity of the leader is greatly enhanced. This enhancement was accomplished in printing the photograph in the darkroom. The right side of the photograph gives some idea of the relative brightness of the stepped leader compared to the return stroke. (Courtesy, Karl Berger, Eidg. Technische Hochschule, Zürich)

viewing portions of the lightning channel end-on. That is, if the lightning channel is coming toward or going away from an observer, the observer will see a greater length of the channel within a given viewing angle than would be the case were the channel perpendicular to his viewing direction. The greater length of channel appears to the eye as a normal channel of greater-than-normal brightness. To account for the beaded effect, the channel must periodically be slanted toward or away from the observer.

(2) Bead lightning is due to the periodic masking of a

Fig. 16.4. Bead lightning as it appears on one frame of a motion picture showing a three-stroke flash striking the top of a depth charge plume. (Courtesy, U.S. Naval Ordnance Laboratory)

normal cloud-to-ground channel by clouds or rain.

(3) A "magnetic pinch-effect" instability may distort the current-carrying channel into a "string of sausages" with the strong light emission coming from the "necked-off" regions of high-current density. To obtain a series of beads this way, the channel must somehow be conditioned periodically as a function of height. Possibly, the periodicity is just an accident of nature.

(4) Bead lightning is a series of ball lightnings (see Chapter 15).

(5) Large-radius sections of the lightning channel have long luminous lifetimes since channels of large radius take longer to cool than channels of small radius. To a first approximation the cooling time is proportional to the square of the radius. If the lightning channel radius were somehow periodically modulated as a function of height, then, as channel luminosity decayed, the channel would take on the appearance of a string of beads. Perhaps this modulation occurs accidentally when the channel consists of a large number of kinks or bends.

To account for the observed long persistence of the beads, those theories which require current flow may be invoked in conjunction with long continuing currents. Perhaps the observed long persistence is due to the eye rather than to the channel.

REFERENCES

16.1.   Young, G. A., A Lightning Strike of an Underwater Explosion Plume, U. S. Naval Ordnance Lab., NOLTR 61-43, March 1962.

A more detailed discussion of bead lightning is given in: *Lightning,* M. A. Uman, Dover Publications, Inc., New York, 1984, Appendix C.
A comprehensive discussion of lightning photography including several streak-camera photographs and references to published photographs of ribbon lightning is given in *Lightning,* Chapter 2.

chapter
seventeen

Has Lightning Any Practical Use?

Energy is neither created nor destroyed. Rather, it is converted from one form to another. For example, in a light bulb the input electrical energy is converted to light and heat. The energy content of the output light and heat equals that of the input electrical energy. No energy is lost. A charged thundercloud contains stored electrical energy derived from the energy of winds and particle motions and from the energy released by condensing water vapor (see Chapter *1*). Some of the stored electrical energy of the cloud is transferred to lightning. We consider now the energy and power available to lightning and the possibility of its controlled use for the benefit of man.

Each cloud-to-ground lightning flash involves tens of coulombs of charge (see Chapter 8) at a voltage of between 100 million and a billion volts. The resultant electrical energy is about 1 to 10 billion watt-seconds per flash. If there are 100 flashes to ground each second over the whole world (see Chapter 7), a maximum value for the total electrical power input to worldwide cloud-to-ground lightning is 1,000 billion watts. By comparison, the total capacity of all electric power generators in the United States in 1970 was about 500 billion watts. It is evident, therefore, that the power and energy available to lightning is appreciable.

Unfortunately, no efficient method of tapping lightning's power and energy is presently available — for two reasons: (1) It is impractical to intercept (for example, with tall towers) any significant number of the world-wide cloud-to-ground flashes. (2) Most of the energy available to the lightning is converted along the lightning channel to thunder, heat, light, and radio waves, leaving only a fraction available at the channel base for immediate use or storage.

The first problem is the more serious. If its total energy were available, a *single* lightning flash would run an ordinary household light bulb for only a few months. It is the high rate of occurrence of world-wide lightning (100 per second) which provides for the high total power and energy levels. A 1000 ft tower in a region of moderate thunderstorm activity is struck by lightning about 10 times a year (see Chapter 6). It would take one hundred thousand such towers, each collecting all the energy from ten strokes per year, to equal the 100 million watts generated by a typical small power station. It is impractical to erect a sufficient number of towers to capture significant quantities of energy from lightning, and no more practical means of intercepting lightning has been proposed.

Ancient man considered lightning the ultimate weapon. Its use and control were generally attributed to the gods. The Norse god Thor hurled lightning from his chariot while riding across the sky. In ancient Greece, Zeus threw the thunderbolts; in Rome, Jupiter. The advent of our modern civilization has not eliminated man's interest in the use of lightning as a weapon. Military research in several countries has recently been concerned with the feasibility of directing lightning and ball lightning at specific targets. While technology is not capable of achieving these goals at present, it would not be surprising if they were attained in the not too distant future. In view of the awesome destructive power of modern weaponry, the military use of lightning or ball lightning would probably be more as a psychological than as a destructive weapon.

Nitrogen comprises about 80% of the atmosphere surrounding the earth, yet it cannot be used directly by the large majority of plants and animals until it is "fixed." "Fixed" nitrogen is nitrogen incorporated in chemical compounds necessary to the chemical processes of life, as opposed to the relatively inert form of nitrogen, the nitrogen molecule, found in the air. Fixing is accomplished by (1) special organisms in the soil and waters, (2) industrial processes, and (3) ionizing atmospheric processes including lightning. Nitrogen which has been fixed in the atmosphere is brought to earth in rain. It has been estimated that of the 100 million tons of nitrogen fixed per year, about 8% is atmospheric, about 30% industrial, and the remainder biological (Ref. 17.1). Estimates of this sort are difficult to make and subject to large errors. The role played by lightning in generating fixed nitrogen is frequently accorded prominence in the popular literature. We can, however, show that lightning is not as important in nitrogen fixing as is generally believed.

About 8 million tons of fixed nitrogen are thought to be brought to earth each year in rain. If we assume that (1) there are 500 lightning discharges (cloud-to-ground and intracloud) occurring each second (see Chapters 7 and 8), (2) each lightning channel is 2 in. in diameter and 4 miles long (see Chapter 10), (3) every nitrogen atom in every channel is "fixed" (a great overestimate), then the total amount of nitrogen fixed by the lightning channels is about 20,000 tons per year. Since 20,000 tons is a very small fraction of 8 million tons, we conclude on theoretical grounds that the lightning channel itself is not an important producer of fixed nitrogen. Experimental evidence to support this conclusion has been given by several investigators (Refs. 17.2 to 17.5) who show that the nitrate and ammonia content in precipitation is not correlated with the amount of lightning activity. In fact, there is roughly the same nitrate and ammonia content in precipitation in winter when there are few thunderstorms as in summer when there are many (Ref. 17.2). What, then, is the source of the atmospheric nitrogen compounds? The answer is not known, but there are two reasonable possibilities: (1) They are generated in the lower regions of the ionosphere (the ionized layer of atmosphere beginning about 40 miles above the earth) and diffuse downward, and/or (2) they are created by corona discharges in all types of precipitating clouds and along the ground under those clouds (Ref. 17.5). If possibility (2) is important, a close relative of lightning has contributed and continues to contribute significantly to the world's fixed nitrogen supply.

The lightning channel generates radio waves (sometimes called atmospherics or sferics) which have been used for practical purposes. Lightning is the strongest terrestrial source of electromagnetic noise in the radio band. The lightning radio noise is strongest near a frequency of 5 kilocycles per second and extends with decreasing intensity up and down the frequency scale as far as has been measured. Everyone is familiar with the lightning static produced on AM radios and the "snow" on TV pictures (particularly on the lower channel numbers) due to lightning radiation. Studies of the lightning electromagnetic noise have led to several practical applications. (1) Measurements made on lightning radio waves can be used to pinpoint the location of the thunderstorm area producing the lightning. Among those who use this information are meteorologists (who study as well as predict the weather) and aircraft pilots. Radiolocation techniques can spot thunderstorms thousands of miles away. (2) Measurements of the properties of lightning radio waves returned from the ionosphere and from ionized layers above the ionosphere have yielded valuable information about the numbers and kinds of charged particles in those regions. (3) Measurements of the properties of lightning radio waves as they propagate along the earth's surface have enabled the electrical resistivity of various portions of the earth to be determined. Similarly, lightning radio waves can be used in some kinds of geophysical prospecting (e.g., in the search for low resistivity ore lying far below the earth's surface).

REFERENCES

17.1.   Delwiche, C. C., The Nitrogen Cycle, *Sci. American, 223* (No. 3), 137-146, September 1970.
17.2.   Viemeister, P. E., Lightning and the Origin of Nitrates Found in Precipitation, *Jour. of Meteorology, 17*, 681-683 (1960)
17.3.   Wetselaar, R., and J. T. Hutton, The Ionic Composition of Rainwater at Katherine, N. T., and its Part in the Cycling of Plant Nutrients, *Australian J. Agr. Res., 14*, 319-329 (1963).
17.4.   Gambell, A. W., and D. W. Fisher, Occurrence of Sulfate and Nitrate in Rainfall, *J. Geophys. Res., 69*, 4203-4210 (1964).
17.5.   Reiter, R., On the Causal Relation between Nitrogen-Oxygen Compounds in the Troposphere and Atmospheric Electricity, *Tellus, 22*, 1-135 (1970).

ADDITIONAL READING

Griffing, G. W., Ozone and Oxides of Nitrogen Production during Thunderstorms, *J. Geophys. Res., 82*, 943-950 (1977).
Chameides, W. L., D. H. Stedman, R. R. Dickerson, D. W. Rusch, and R. J. Cicerone, NO_x Production in Lightning, *J. Atmosph. Sci., 34*, 143-149 (1977).
Hill, R. D., R. G. Rinker, and A. Coucourinos, Nitrous Oxide Production by Lightning, *J. Geophys. Res., 89*, 1411-1421 (1984).
Levine, J. S., T. R. Augustsson, I. C. Anderson, and J. M. Hoell, Jr., Tropospheric Sources of NO_x: Lightning and Biology, *Atmospheric Environment, 18*, 1797-1804 (1984).

chapter eighteen

What Would Happen If Lightning Were Eliminated?

First, it is appropriate to point out that it may well have been lightning in the primordial soup covering the earth several billion years ago that produced the complex molecules from which life eventually evolved. Laboratory experiments have shown that electrical discharges in what is believed to be the constituents of the primordial atmosphere can create the necessary molecules. Thus, we may be indebted to lightning for the presence of life on earth.

Still, lightning is primarily troublesome rather than helpful, as the contents of this book attest, and except for some of the benefits mentioned in the previous chapter it might seem advantageous to eliminate it altogether. Exactly what would happen if lightning were eliminated is not known, but clearly the electrical balance of the atmosphere would have to change as we shall see in the following paragraphs. If and

151

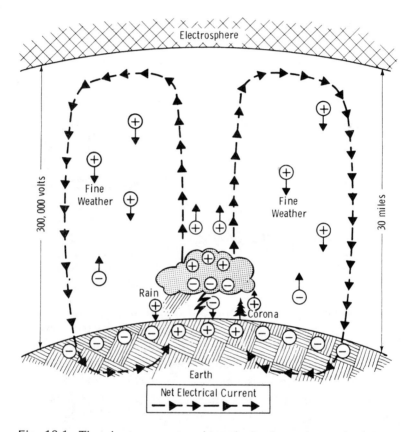

Fig. 18.1. Thunderstorms act as batteries to keep the earth charged negatively and the atmosphere charged positively. Atmospheric electrical currents flow downward in fine weather and upward in thunderstorms. Thunderstorms deliver charge to the earth by lightning, rain, and corona discharges.

how the change in electrical balance would change the weather is not known. Prudence, however, would dictate extreme caution in tampering with lightning.

The electrical resistivity of the atmosphere decreases with height. From the point of view of atmospheric electricity, the resistivity is sufficiently low at an altitude of about 30 miles that the voltage does not vary much above that point. The region beginning at about 30 miles and extending upward is called the *electrosphere*. (The ionosphere, which reflects lightning radio waves downward, begins, as noted in Chapter 17, at about 40 miles and is even a better conductor than the bottom of the electrosphere.) The voltage between the earth and the electrosphere in regions of fine weather is about 300,000 volts. To maintain this voltage the earth has a negative charge of about a million coulombs on its surface and an equal net positive charge is distributed throughout the atmosphere. Measurements have shown that the negative charge on the earth remains roughly constant with time. At first glance, this fact is difficult to understand since the charge on the earth is continuously leaking off into the conducting atmosphere. In fact, calculations show that, if the earth's charge were not being continuously re-supplied, the charge on the earth would disappear in less than an hour.

The earth is recharged by thunderstorms. Fig. 18.1 shows how the electrical balance of the atmosphere is maintained. Thunderstorms deliver a net negative charge to the earth as a result of the sum of the effects of the following processes: (1) negative charge carried from cloud to earth by lightning, (2) positive charge carried from cloud to ground by rain, and (3) positive charge carried upward (the equivalent of negative charge carried downward) through the air beneath and above a thunderstorm, the source of the positive charge being corona discharge off grass, trees, and other objects with sharp points on the ground beneath the thunderstorms. The total current flowing beneath all thunderstorms in progress

Fig. 18.2. Lightning over Lake Chiem in Bavaria, Germany. (Courtesy, Agfa-Gavaert AG)

Fig. 18.3. Lightning over the Arizona desert. (Courtesy, J. Rodney Hastings, University of Arizona)

Fig. 18.4. A lightning flash exhibiting multiple ground points and the ribbon effect discussed in Chapter 16. (Courtesy, George Marcek, Catalina High School, Tucson)

Fig. 18.5. Lightning initiated by an upward moving leader from a tower on Mt. San Salvatore near Lugano, Switzerland. Photographs of other discharges to the tower are shown in Fig. 6.1a, b. (Courtesy, Richard E. Orville, State University of New York at Albany)

throughout the world at any given time is thought to be about 2000 amps, and is in such a direction as to charge the earth negatively. An approximately equal and opposite current flows in regions of fine weather. The result is that the net negative charge on the earth and the equal and opposite net positive charge in the atmosphere remain approximately constant.

Finally, if all lightning were eliminated, we earthly viewers would be deprived of one of the most spectacular visual displays that Nature has to offer. In its variety and brilliance lightning puts man-made fireworks to shame, as evidenced by Figs. 18.2 to 18.5. Watching lightning is fun; one of the aims of this book has been to make it more so. Many confirmed lightning watchers even enjoy listening to the variety of sounds that comprise thunder.

REFERENCES

The details of and references to the bulk of the material found in this chapter are to be found in *Atmospheric Electricity*, J. A. Chalmers, 2nd Edition, Pergamon Press, New York, 1967, pp. 33-35, 292-308.

Martin A. Uman received the B.S.E. degree in electrical engineering in 1959 from Princeton University, where he was elected to Phi Beta Kappa. After receiving his Ph.D. from Princeton University in 1961, he took a post as Associate Professor of Electrical Engineering at the University of Arizona, where he became interested in lightning research, particularly in lightning spectroscopy. In 1965, Dr. Uman moved to the Westinghouse Research Laboratories in Pittsburgh, where he continued his  lightning research and, in addition, studied the electrical, optical, and acoustic properties of long laboratory sparks. He has been Professor of Electrical Engineering at the University of Florida since January 1972, where his primary research interest has been in the electromagnetic radiation produced by lightning. In 1976, Dr. Uman and Dr. E. P. Krider, Professor of Atmospheric Sciences at the University of Arizona, formed Lightning Location and Protection, Inc. (LLP), a small company that specializes in consulting on unusual lightning problems and manufacturing lightning detection equipment. Dr. Uman served as president of LLP from 1976 until 1983. Dr. Uman is the author of three books—*Lightning, All About Lightning,* and *Introduction to Plasma Physics*—and approximately 85 journal articles, and has served as Associate Editor of the *Journal of Geophysical Research.* He is a member of the International Commission on Atmospheric Electricity, the American Geophysical Union, and the IEEE, and is a fellow of the American Meteorological Society.

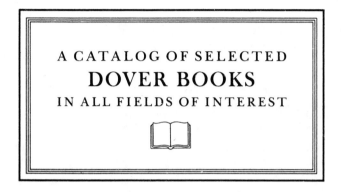

A CATALOG OF SELECTED
DOVER BOOKS
IN ALL FIELDS OF INTEREST

A CATALOG OF SELECTED DOVER BOOKS IN ALL FIELDS OF INTEREST

DRAWINGS OF REMBRANDT, edited by Seymour Slive. Updated Lippmann, Hofstede de Groot edition, with definitive scholarly apparatus. All portraits, biblical sketches, landscapes, nudes. Oriental figures, classical studies, together with selection of work by followers. 550 illustrations. Total of 630pp. 9⅛ × 12¼.
21485-0, 21486-9 Pa., Two-vol. set $25.00

GHOST AND HORROR STORIES OF AMBROSE BIERCE, Ambrose Bierce. 24 tales vividly imagined, strangely prophetic, and decades ahead of their time in technical skill: "The Damned Thing," "An Inhabitant of Carcosa," "The Eyes of the Panther," "Moxon's Master," and 20 more. 199pp. 5⅜ × 8½. 20767-6 Pa. $3.95

ETHICAL WRITINGS OF MAIMONIDES, Maimonides. Most significant ethical works of great medieval sage, newly translated for utmost precision, readability. Laws Concerning Character Traits, Eight Chapters, more. 192pp. 5⅜ × 8½.
24522-5 Pa. $4.50

THE EXPLORATION OF THE COLORADO RIVER AND ITS CANYONS, J. W. Powell. Full text of Powell's 1,000-mile expedition down the fabled Colorado in 1869. Superb account of terrain, geology, vegetation, Indians, famine, mutiny, treacherous rapids, mighty canyons, during exploration of last unknown part of continental U.S. 400pp. 5⅜ × 8½. 20094-9 Pa. $6.95

HISTORY OF PHILOSOPHY, Julián Marías. Clearest one-volume history on the market. Every major philosopher and dozens of others, to Existentialism and later. 505pp. 5⅜ × 8½. 21739-6 Pa. $8.50

ALL ABOUT LIGHTNING, Martin A. Uman. Highly readable non-technical survey of nature and causes of lightning, thunderstorms, ball lightning, St. Elmo's Fire, much more. Illustrated. 192pp. 5⅜ × 8½. 25237-X Pa. $5.95

SAILING ALONE AROUND THE WORLD, Captain Joshua Slocum. First man to sail around the world, alone, in small boat. One of great feats of seamanship told in delightful manner. 67 illustrations. 294pp. 5⅜ × 8½. 20326-3 Pa. $4.95

LETTERS AND NOTES ON THE MANNERS, CUSTOMS AND CONDITIONS OF THE NORTH AMERICAN INDIANS, George Catlin. Classic account of life among Plains Indians: ceremonies, hunt, warfare, etc. 312 plates. 572pp. of text. 6⅛ × 9¼. 22118-0, 22119-9 Pa. Two-vol. set $15.90

ALASKA: The Harriman Expedition, 1899, John Burroughs, John Muir, et al. Informative, engrossing accounts of two-month, 9,000-mile expedition. Native peoples, wildlife, forests, geography, salmon industry, glaciers, more. Profusely illustrated. 240 black-and-white line drawings. 124 black-and-white photographs. 3 maps. Index. 576pp. 5⅜ × 8½. 25109-8 Pa. $11.95

AMERICAN CLIPPER SHIPS: 1833–1858, Octavius T. Howe & Frederick C. Matthews. Fully-illustrated, encyclopedic review of 352 clipper ships from the period of America's greatest maritime supremacy. Introduction. 109 halftones. 5 black-and-white line illustrations. Index. Total of 928pp. 5⅜ × 8½.
25115-2, 25116-0 Pa., Two-vol. set $17.90

TOWARDS A NEW ARCHITECTURE, Le Corbusier. Pioneering manifesto by great architect, near legendary founder of "International School." Technical and aesthetic theories, views on industry, economics, relation of form to function, "mass-production spirit," much more. Profusely illustrated. Unabridged translation of 13th French edition. Introduction by Frederick Etchells. 320pp. 6⅛ × 9¼. (Available in U.S. only)
25023-7 Pa. $8.95

THE BOOK OF KELLS, edited by Blanche Cirker. Inexpensive collection of 32 full-color, full-page plates from the greatest illuminated manuscript of the Middle Ages, painstakingly reproduced from rare facsimile edition. Publisher's Note. Captions. 32pp. 9⅜ × 12¼.
24345-1 Pa. $4.95

BEST SCIENCE FICTION STORIES OF H. G. WELLS, H. G. Wells. Full novel *The Invisible Man*, plus 17 short stories: "The Crystal Egg," "Aepyornis Island," "The Strange Orchid," etc. 303pp. 5⅜ × 8½. (Available in U.S. only)
21531-8 Pa. $4.95

AMERICAN SAILING SHIPS: Their Plans and History, Charles G. Davis. Photos, construction details of schooners, frigates, clippers, other sailcraft of 18th to early 20th centuries—plus entertaining discourse on design, rigging, nautical lore, much more. 137 black-and-white illustrations. 240pp. 6⅛ × 9¼.
24658-2 Pa. $5.95

ENTERTAINING MATHEMATICAL PUZZLES, Martin Gardner. Selection of author's favorite conundrums involving arithmetic, money, speed, etc., with lively commentary. Complete solutions. 112pp. 5⅜ × 8½.
25211-6 Pa. $2.95

THE WILL TO BELIEVE, HUMAN IMMORTALITY, William James. Two books bound together. Effect of irrational on logical, and arguments for human immortality. 402pp. 5⅜ × 8½.
20291-7 Pa. $7.50

THE HAUNTED MONASTERY and THE CHINESE MAZE MURDERS, Robert Van Gulik. 2 full novels by Van Gulik continue adventures of Judge Dee and his companions. An evil Taoist monastery, seemingly supernatural events; overgrown topiary maze that hides strange crimes. Set in 7th-century China. 27 illustrations. 328pp. 5⅜ × 8½.
23502-5 Pa. $5.95

CELEBRATED CASES OF JUDGE DEE (DEE GOONG AN), translated by Robert Van Gulik. Authentic 18th-century Chinese detective novel; Dee and associates solve three interlocked cases. Led to Van Gulik's own stories with same characters. Extensive introduction. 9 illustrations. 237pp. 5⅜ × 8½.
23337-5 Pa. $4.95

Prices subject to change without notice.
Available at your book dealer or write for free catalog to Dept. GI, Dover Publications, Inc., 31 East 2nd St., Mineola, N.Y. 11501. Dover publishes more than 175 books each year on science, elementary and advanced mathematics, biology, music, art, literary history, social sciences and other areas.